The Housing Downturn

Picking Up the Pieces

A Guide for Estate Agents and Developers

Graham Norwood

2009

A division of Reed Business Information

Estates Gazette
1 Procter Street, London WC1V 6EU

ISBN 978-07282-0570-3

Typeset in Palatino 10/12 by Amy Boyle, Rochester
Cover design by Rebecca Caro
Printed by Bell & Bain, Glasgow

Contents

Preface

There are looks of deep concern on the faces of estate agents and developers in the UK. They have had an extraordinarily bad time with worse to come, it seems. However, talk with them and, sooner rather than later, a smile may appear.

The reason? Well, the housing market's runaway train, with its prices soaring from 1993 until 2007, was finally stopped; but not by them.

A few tabloids and their readers may still think agents and developers are the devil incarnate but most of the country — no, make that the world — blame the banks and not the property industry for the slump in house prices and the broader economic malaise gripping the globe.

If one thing has become a given as we hit the second decade of the 21st century, it is that the banks and other lenders have been too lax, too cavalier, too optimistic and almost certainly insufficiently honest with investors, borrowers and each other. There is an absolute imperative for them now to restructure their balance sheets to fund more or even all of their loans from deposits and not by borrowing on wholesale markets, which is where the roots of the credit crunch lie.

What is more, while the banks take the blame, it appears that the housing market has — we think — escaped suffering a wholesale crash. Many would use the "C" word but I believe it is wrong to do so. This book does not skimp on revealing how a few areas have seen prices collapse by as much as 50%, but I carefully avoid using the term "crash" throughout these pages for two reasons.

The first is because I think the word is inaccurate in describing what has happened to the residential market when seen over the long-term.

If an average 30% or even 35% price fall for homes constitutes a "crash", then what happened to oil prices (down from $145 per barrel to $40 in just two months in late 2008) would be described as a cataclysmic collapse. But we know the longer-term trend for oil is upwards, so just about no one uses the term "crash" for a relatively short-term slump in that commodity's price.

In 2008, for example, UK house prices fell 12% to 18% depending on which index you use or believe. Over the calendar year, oil fell by 58% (and we know it fell far more if you do not look at the whole year and simply look at its highest point, in the early autumn, and its lowest point, just before Christmas). Meanwhile, Sterling's value on world markets fell 24% in the calendar year and the FTSE-100 share index dropped 31%. Against those measures, too, would you say that house prices crashed? No.

Even now I believe that residential prices, like oil, will rise in value over the long-term and within the next decade prices will surpass those seen at the recent peak, 2007. Classic capitalism may have its huge weaknesses but the shortage of homes within the UK means that a lack of supply compared to demand will push prices up, notwithstanding the over-supply that exists in some areas of the country.

The second reason I do not use the word "crash" is because a few of my journalistic colleagues have debased its value.

Most business, finance, property and industry writers have a decent enough understanding of the housing market and so have usually avoided the "C" word in recent years. However, a few hacks have taken the easy way out and extrapolated one adverse house price index or one example of a distress sale into a permanent trend.

Inevitably, that meant "crash" appeared in a small number of headlines, and equally inevitably that meant some less sophisticated property insiders started blaming the press rather than anyone else — even the banks.

So if you have gone as far as to buy this book, the likelihood is we can agree on two things at this early stage: the UK housing market suffered an unruly downturn but not a crash, and the root cause lies with the banks rather than the property industry.

But before we all engage in a group hug, think about this.

Many agents and developers fell by the wayside in the downturn years following late 2007. But did they have to? Were they as well prepared and appropriately structured as they could have been, ahead of what was an inevitable downturn?

Of those agents and developers who struggled through the slump, are there lessons they could teach the rest of us regarding their business models, their attitudes and their skills?

I think so, because it would surely be a mistake to simply wait for an upturn and assume all will then revert to the same world that we inhabited between 1993 and 2007.

That world will clearly not be the same, and estate agents and residential builders may well suffer more and for longer if they do not use the current malaise as an opportunity to reconsider and reshape how they work.

This book does not pretend to have complete answers, but is perhaps most useful in raising questions — the classic role of the journalist during the ages.

My position is one which has many advantages.

As a freelance journalist writing solely about residential property for more than a decade now, I have chronicled the growth in the industry in recent times. It seemed a very basic business back in the mid-1990s. It was before the biggest surge in buy-to-let; before large-scale international property ownership and investment; before the internet revolutionised sales, searches and buyer awareness; before Home Information Packs; before the rise (and fall?) of "city centre living".

But the biggest advantage is that although I have an understanding of the industry I also have an outsider's view, weighed down by neither the workload of the agent nor of the developer. My position also avoids having to be unnecessarily respectful of industry processes just because "that's how they've always been done".

So that is why it appears clear to me that many agents and developers can do more to protect themselves from the next inevitable downturn, whenever that may be.

If that protection had been in place five years ago, perhaps more industry players would have spotted the signals that were missed and which we try to identify in Chapter one. Then, perhaps more agents and builders could have survived the fate that befell them, chronicled in later chapters of this book.

What are those forms of protection? What could the property industry have used: new technology, taken tips from other industries to better shelter itself and perhaps even build business in the future? We look at all that, too, before we assess what the future might hold — good and bad.

The "we" in question are the scores of property analysts, estate agents, developers and other industry insiders who kindly gave me their time and views since the start of the downturn. My thanks are due to them and also to Alison Bird, Commissioning Editor at EG Books, and my ever-patient wife Helen Crossfield.

The best parts of this book are down to these people: the rest is down to me.

Graham Norwood
March 2009

1 What were you doing on 9 August 2007?

It is pointless to try to identify the "start" of the credit crunch.

Some observers say it was in the spring of 2007, when the US lender New Century Finance filed for Chapter 11 bankruptcy protection — that heralded the start of global public awareness of a problem with something called sub-prime mortgages.

Others say it was in July that year, when Bear Stearns (remember that company?) told investors in two of its hedge funds that they would not be compensated, because other banks refused to lend Bear Stearns any money for a bail-out. That was the incident which signalled the reluctance of banks to lend to each other.

Still others say the start of the crunch was not one event but a protracted process dating from the late 1990s. That was when cheap loans, some offered by those we can now safely describe as cheap-suited wide boys, became commonplace across most of the major economies and especially in the UK and the US.

In any case, the exact start date is not important. In the early months of 2007, few in the British property industry had any knowledge of arcane Wall Street lending processes or the way in which the major US financial institutions bundled up debts and sold them like a cut-price contagion around the world.

Even a year later, as 2008 dawned, the property industry had no idea what was going to happen.

Most commentators were optimistic, buoyed by a Bank of England interest rate cut in December 2007. They were anxious to silence the occasional siren voice predicting doom and gloom.

"We expect economic conditions to be more difficult for the housing market, but we do not expect a recession. Interest rates are on the way down and there is the ongoing issue of under-supply. Consequently the underlying fundamentals are perhaps more positive than recent swings in sentiment might suggest" was how the Nationwide Building Society's Chief Economist, Fionnualla Earley, saw the year ahead in January 2008.

Estate agency King Sturge predicted a 4% rise in average prices across the UK and 6% in London; Savills predicted a 3% rise, as did Knight Frank; property consultancy Hometrack plumped for 1%.

If only they, and the rest of us, paid attention on 9 August 2007.

For that was the day the credit crunch came to our backyard — or at least to Europe's backyard — even if we did not recognise the problem at the time. On that day BNP Paribas, a French investment bank, told its investors that they could not withdraw money from two of its funds because of what it called "a complete evaporation of liquidity" in the market place.

Suddenly what was happening across the Atlantic was happening across the Channel. The European Central Bank acted quickly. Within a week it put over £130 billion into the banking market — the first of hundreds of such interventions by central banks on all five continents in the following 18 months.

It is probably a safe bet that none of us realised then what those actions meant, so the odds are that you do not actually remember what you were doing on 9 August 2007. But that is exactly the point of this chapter; we seem to have missed the signals.

On that day we were more occupied with the Land Registry's news that English and Welsh house prices were still rising by an average of 9.4% (the highest rate for over a year and a level which has not been matched since, of course).

Or perhaps as a country the UK was too busy completing its routine daily quota of business. Back then it was, believe it or not, 5,000 house sales and 160 buy-to-let mortgages being processed every working day.

We in the property business were, of course, getting on with stuff. Stuff that fuelled the property market, put roofs over people's heads, helped many realise their dreams, and paid our own salaries and mortgages.

But why did we not notice what was going on, even as late as mid-2007?

Why were developers not restructuring to reduce exposure to a burgeoning buy-to-let sector? Why were agents still expanding their branch empires — in August 2007 an estimated 60 new branches opened in England alone, a sign of an industry believing it was growing, not accelerating towards an abyss?

Liam Bailey, head of residential research at Knight Frank, has been a leading residential analyst for five years and believes the industry made three errors.

First, it assumed the long-term low level of house building throughout the UK would keep demand and therefore prices high, irrespective of other factors.

Second, he says people were focussed on bricks and mortar and did not anticipate a fast-moving mortgage drought.

Third, too many analysts failed to link the US sub-prime crisis with the UK. "Too many commentators were still too UK-centric in their world view. The global nature of the bubble was missed" he suggests.

"The only publication to really forecast it was the Economist, which had a world house price index back in 2004 saying the market was deflating and would eventually burst" says Bailey.

He says the real estate boom had enjoyed so many false sunsets that no one recognised the real one:

> In early 2003 the market slowed down and we all thought it was the start of a fall. But by the summer of that year, it was moving up again. It did the same thing at the end of 2004 and into the start of 2005 — prices stagnated and demand dropped, only for it all to gather momentum again by the summer. So when we saw things slowing in late 2007, we probably didn't react as quickly as we could.

Rival analyst Jenneit Siebrits of CB Richard Ellis agrees that it was the international inter-relatedness that caught the UK industry on the hop:

> It was not so much that signals were missed, but the global climate changed so rapidly that analysis of key indicators could not have predicted where we are now. The sub-prime crisis affected almost every market and the extent to which banks had unwittingly entangled themselves was unclear until Bear Sterns collapsed. The fall of Lehman Brothers then confirmed the inevitable.

Nick Goble is an estate agent who entered the industry at the time of the 1990s recession and now is the franchise holder for Winkworth's agency across part of south London, he stated:

> I don't think we actually missed much over the last 10 years. We saw a strong economy, with what we thought was sustainable growth. We all knew that the loan to value ratio and mortgages were increasing. This, however, did not concern us in general as interest rates were historically low ...
>
> I remember during the last recession interest rates were at 15% and inflation was at 9%. We all thought the [more recent] increase in the lending barriers were affordable.

So what was different now? Goble says:

> When I first started as an agent there was not really a global economy on the scale we have today. Communication was more limited, which I think was critical in maintaining confidence. Too much information is not necessarily a positive thing. Ignorance was always bliss, as they said.

Goble's view is typical of many highly successful estate agents, especially those who worked mainly or solely in sales. Douglas & Gordon's Ed Mead — another agent who rose to prominence in the 1995–2007 boom, writing national newspaper columns and even being a subject in a fly-on-the-wall TV documentary series — says the signals were not so much missed, as wrongly interpreted:

> The signs were in the basics. We were, for the first time ever, in a long term low interest environment. This skewed people's sense of perspective in the same way it skewed the banks' perspective. That's why income-value multiples and lending criteria became so dangerous.

Peter Rollings of Marsh & Parsons, and formerly Managing Director of Foxtons, says:

> I'm not sure that one can look back and say that we missed the dangers of 'leveraged securitised debt'.
>
> It seems from reading what the commentators are saying; even the people dealing with it didn't understand it so what chance has the average man in the street got?

Rollings says it is wrong for the industry, and perhaps especially those outside it, to beat itself up and blame estate agents for failing to act:

> It's easy to do and common place. [But] I think one must not lose sight of the fact that it is the job of the agent to get the most amount of money that

> anyone will pay for a property. Where that particular purchaser gets their money from and how much they are borrowing and at what rate and level, is frankly immaterial to the job.

And yet, and yet ... those signals were there. Rollings adds:

> With the benefit of hindsight, one could see that there was something wrong when buyers were coming to us having arranged their mortgage sometimes at six or seven times their annual salary. In general, if one is borrowing 20% more than the property is worth at today's prices, it is in effect saying that this is a one way bet. We all know from the last recession that this is not true.

This unsustainable approach to lending and buying — not to mention prices — reads now to some people, just a few years on, as an example of poor judgment and complacency by the industry as a whole. But back then, we all just thought these phenomena were part of the price to be paid for a buoyant property market.

But whilst macro-economics were easy to misread or overlook by the property industry, which has never been run by economists, there were in retrospect many other signs within our own industry that should have started the alarm.

A few obvious examples are discussed in the following sections.

Missed signal 1: far too many new apartments ... and far too expensive

The headlines in October 2008 were dominated by financial gloom as institutions fell and banks were nationalised. But one piece of housing market news — equally gloomy — penetrated the front pages.

It was that city centre apartments across the UK had seen the largest falls in value since the start of the housing slump, according to *www.mouseprice.com*, a little-known website with a surprisingly sophisticated line in statistical analysis. It said flats in the Birmingham Canal area had seen the greatest falls in value, while flats in the Deansgate quarter of central Manchester were the second fastest fallers. Almost all large city centres featured elsewhere in the firm's league table of declining prices.

This was not yet another index saying, in terms so broad they were almost impossible to pin down, that average prices across the whole country were dropping. Instead, Mouseprice said the falls were not actually linked to the desirability of an area at all and instead were most prominent in those locations where a large number of flats had been built over recent years.

"Price falls have not discriminated according to how much a property cost in the first place or how desirable an area was" said a Mouseprice spokesman. "A clear theme has emerged with the top 10 areas dominated by northern city-centre regeneration sites. These areas attracted the most buy-to-let investors in the boom and have recently suffered due to supply outstripping demand. But falls are not limited to regeneration areas and have spread to unexpected locations."

There should have been no surprise there, and we should have spotted it earlier.

Northern city centres had become the whipping boy of the property market during 2008, and with good reason. If only they had been under similar scrutiny in, say, 2005 then some of the worst effects of the downturn may have been avoided

Data from Knight Frank — the estate agency that invented the catchphrase "city centre living" and had become the most vocal advocate of buy-to-let in northern city centres for five years after the Millennium — show that in 1995 just 20,000 people lived in the combined city centres of Manchester, Liverpool, Newcastle, Leeds and Sheffield.

By 2005 that figure had become 47,000 with annual inflation for those properties running as high as 21% in selected areas. But the pace and volume of development continued even beyond that dramatic scale of repopulation, spurred on by the property industry's short-term profit-seeking and, of course, by the endless grind of government planning directives seeking ever more dense development.

Leeds city centre had 3,655 new flats built by mid-2005, with another 2,408 under construction, 4,808 seeking planning permission and another 4,963 in long-term plans.

In Sheffield, 2,445 had been built with 1,262 under construction, 2,112 with planning and another 2,484 in the long-term. In Manchester there had been 5,630 built, 2,377 under construction, 2,468 with planning and 3,000 in long-term proposals. Over in Liverpool some 3,523 had been built by mid-2005 with 3,049 under construction, 1,548 with planning and 2,562 in long-term proposals.

Newcastle, where the local authority was starting to reject planning applications of flats and suffered orchestrated abuse from developers for doing so, some 2,421 had been built but only 480 were under construction. Yet 1,228 were awaiting planning consent and another 2,419 were sitting on paper for the long-term.

Then there was Nottingham, Glasgow, Edinburgh, Bristol, and even small towns like Slough and Basingstoke which were all being swamped with schemes of two-bed, two-bath flats. The early years of repopulation had ended by 2006 so flats completed, under construction or at planning stage from that point on had no guaranteed buyer (landlord or owner occupier) and was directly contributing to what became the housing market slowdown.

Some siren voices were speaking out against this endless glut.

Back in 2005, Andrew Wells of the Leeds-based auction house Allsops — a long-term critic of the flat-building obsession that gripped his city and many others shortly after the Millennium — noted that literally thousands of landlords' resale apartments were vying for buyers with the seemingly endless production line of new flats.

Again in 2005, the developer CALA Homes sold 916 new homes of which 63% were apartments — a high figure, but nonetheless lower than the 67% of 2004. "We're aiming to reduce this further to 50% over the next three years" promised CALA's Development Director Robert Millar, who added:

> The market for apartments has weakened significantly — with the city centre apartment market reaching saturation point — as the proportion of apartments being built soared to unprecedented levels in a very short space of time, from 36% two years ago [in 2003] to 55% in 2005.

That seems a prescient view today but back then Millar was in a minority within the property industry. There were many other developers and agents, often with substantial public relations budgets and scant regard for honest analysis, who poured scorn on the few who noticed there were just too many flats.

Jonathan Morgan, a Leeds estate agent selling city centre apartments and who has forged alliances with other agents in the city, remains unapologetic about it all. Back in 2005 it was the same story as in the pages of Estates Gazette he lambasted "sensational articles in national newspapers". By 2008 he had set up a campaign to counter what he believed was negative publicity about the over-supply in the city, and this time he even regarded *Estates Gazette* as a source of the sensational articles.

"Leeds, unlike many of its peer cities, has carried out academic research on the city living market which has in turn sparked a grown up debate about the pace and nature of supply. The findings of this report are increasingly referenced by developers when appraising the viability of land opportunities" he says.

By the end of 2008 things were looking rather different.

It was easy to find former buyers of Leeds city centre apartments who were by then selling them for as little as 60% of the price they paid, while some of the early advocates of the "city centre living" idyll were having troubles of their own.

Simon Morris, a former Director of Leeds United and owner of a portfolio of properties reported to include 500 Leeds apartments, put his firm SRM Holdings into administration late in 2008. He was the developer of a £36m residential scheme in Leeds, had been injured in a reported drive-by shooting, and had been named in a BBC TV *Panorama* programme about the controversial sale of student flats. Landlords who had bought buy-to-let properties from one of his firms claimed they had paid too much and many said they had properties repossessed after failing to cover their mortgage repayments.

But the glut of apartments, although greatest in the northern city centres, was by no means restricted to that region.

Figures released in 2008 — then looking back on 2007 as the final "real" year of house building before the slowdown hit — showed that the principal housing type being started in England, Wales and Northern Ireland was, of course, "the apartment".

According to the National House Building Council's (NHBC) statistics, flats and maisonettes made up 51% of new homes started in England during the third quarter of 2007 — almost double the combined percentage of semi-detached and detached homes.

In Wales, NHBC statistics show that the percentage of flats and maisonettes started in the third quarter of 2007 rose from 34% in the second quarter to 46%, over double the 21% figure for detached homes.

In Northern Ireland, the entire year of 2007 experienced a significant rise in the number of flats and maisonettes when compared to 2006. NHBC statistics showed that flats and maisonettes constituted 29% of all new homes started in the year — five percentage points higher than detached homes started and nine percentage points higher than semi-detached homes started during the same period.

2007's third quarter figures for Scotland show flats and maisonettes rivalling detached houses — 37% compared with 39%.

Knight Frank's Liam Bailey says there was a kind of blindness at the time, believing apartments were the right way to go:

> There was so much government spin around targets, we blindly thought gluts of one and two bedroom flats were partly, perhaps significantly, the answer to our problems. What we didn't consider was that the apartments were being built above pubs and clubs in city centres, and the growing numbers of single-person households we were being told about were actually little old ladies outliving their male partners.

Just a few years on, this all seems absurd. It reads like a disaster waiting to happen, and happen it did. Why did we miss it?

Missed signal 2: comparisons with the early 1990s

Too often throughout late 2007 until early 2009, housing commentators would say that the downturn must be less severe than in the early 1990s because the wider British (and indeed global) economy was in better shape now than then.

But that was a falsely complacent comparison — the British economy was indeed better in 2008 than in 1990, but so what? The differences that counted were to do with financing, not the headline inflation, unemployment or productivity rates.

"Turnover rates in the housing market have fallen to historic lows, even below the levels in the 1990s when the economic conditions were worse than they are today. At the trough of the market in quarter four 1990, interest rates were at 14% and there were almost double the number of unemployment claimants, yet a greater proportion of owner occupiers were taking out mortgages to move house" notes Fionnuala Earley, Chief Economist of the Nationwide Building Society.

"The significant difference today is the financial market shock which has led to the severe tightening of credit. In quarter four 1990, 60% of first time buyers were taking out loans with loans-to-value above 90%, today the equivalent proportion is 14%. While this may reflect less desire on the part of borrowers to borrow at high LTV, especially given its higher cost, it also implies that part of the reduction in turnover today is likely to be due to the availability of finance at higher LTV" she says.

Missed signal 3: Countrywide's clever move

The estate agency group Countrywide plc has always been a big indicator of the market, as it should be — in 2007 it handled around one in 12 of all residential sales and quite a few lettings, mostly at the mid- to low-end of the market.

In May 2007, Countrywide plc was acquired by the private equity finance group Apollo Management LP, which took the group out of public ownership. It closed many offices — after all, it had literally dozens of high street brand names operating side by side, so the scope for rationalisation was obvious. But while it made it clear that it expected a downturn, perhaps the biggest sign that we missed was that it toggled.

Toggled? Well, yes. I cannot explain this arcane financial procedure better than Robert Peston, the BBC's Business Editor who became a legend (not to mention a harbinger of doom) during the credit crunch. He put it like this in his BBC Online blog early in 2008:

> Britain's largest estate agent, Countrywide, has toggled. Or to translate, it has stopped paying interest in cash on £100m of debt and is instead rolling it up.
>
> It's a sign of the tsunami that's hit Britain's estate agents that Countrywide — which has about 8% of the British residential market — has decided to conserve as much cash as it can.
>
> However, under the extraordinarily favourable borrowing terms negotiated by Countrywide's private-equity owner, Apollo of the US, when buying this business for more than £1bn a year ago, Countrywide is quite within its rights to stop paying this interest in cash — and there's nothing its lenders can do but wince.
>
> In fact clever old Countrywide has also just drawn down a further £100m odd from a bunch of banks under a revolving credit facility it negotiated at the time of the buyout. It's done this not because it is strapped for cash right now but as insurance against the risk that its cash-flow from operations will be insufficient to pay its residual interest over the coming few years.
>
> Yes, you read that correctly. A bunch of banks have lent to Countrywide to allow it to make interest payments to another bunch of lenders. This would be financial commonsense only in Wonderland.
>
> And I have to ask what was on the mind of banks and other financial institutions when they financed Countrywide's buyout.
>
> It was a deal done at the very peak of irrational exuberance about private equity last summer. The buyer, Apollo, obtained the most

> astonishingly favourable terms from lenders — including the absence of the normal covenants that allow lenders to take control of a business when the going gets tough.
>
> In the case of Countrywide, it seems there would literally have to be Armageddon for the lenders to have any real power. Apparently a fall in the volume of house sales this year of over a third, which is what the Council of Mortgage lenders expects and is in line with Countrywide's own estimates, does not represent Armageddon.
>
> The boss of Countrywide, Grenville Turner, tells me that — having drawn on the revolving facility and toggled the £100m floating rate note — Countrywide has enough readies to pay interest on its residual £640m of cash-interest-paying debt for three years, even if the housing market remains flat for as long as that (which he says he doesn't expect).
>
> As for the thousands of agents employed in Countrywide's 1,100 residential offices, they have reason to take comfort from bankers' pain.
>
> If it weren't for the way that Apollo screwed this mind-boggling deal out of the lenders — if Countrywide hadn't been able to suspend cash-interest payments and borrow to pay interest — the prospects for this business and their jobs would be a good deal worse (though in view of the mess in their market, that's probably not a reason to be too cheerful).

Whether or not one understands every intricacy in this characteristically labyrinthine financial process, the point is clear.

News of this deal began to circulate around the accountancy divisions in Britain's estate agents as early as January 2008 — already too late to fundamentally avoid the slowdown but still with time to see that if the UK's largest agent was worried, so we all should be too.

It remains to be seen whether this deft move will be enough to save Countrywide in the long-term. In late 2008, some months after the toggling procedure, the credit ratings agency Standard & Poor's downgraded Countrywide's credit status, saying it could run out of money within 12 months. This followed the group showing operating losses of £29m in the first nine months of 2009 and the fact that transactions had fallen by more than the one-third that its 2007 prediction had been based on.

Missed signal 4: lenders' systematic down-valuing

Amongst the first people to notice the slowdown were, of course, the mortgage lenders. They started to see what became known amongst many lenders as "cancer corners" — blocks of new apartments being built, mostly in city centres. This led to a rapid and widespread decision by lenders to down-value new homes.

Back in November 2007, the Nationwide Building Society became the first lender to intentionally do this. At that time the Society applied it only to all new apartments, but eight months later the policy was extended by include new-build houses too, as they appreciated the breadth of the slowdown.

The policy became known to the Nationwide's panel of valuers, who received a string of letters telling them they should value new flats according to their estimated re-sale price, excluding any new-build premium.

Ironically, this made it harder for some purchases to go ahead and, some went on to argue, potentially distorted the Nationwide's own house price index. Some months later the Council of Mortgage Lenders (CML) and the Royal Institution of Chartered Surveyors (RICS) — which authored guidelines to help valuers make impartial judgments, introduced in September 2008 — admitted that several lenders were doing the same, automatically dropping to second hand values.

"It left buyers in the lurch. They agreed a deal in principle with a builder, possibly already including a price cut. Then the buyer applies for a mortgage but the lender gets a valuation at an automatically lower price. So unless the buyer made up the shortfall, the deal collapsed" says Lee Holland of Town & Country Mortgages.

"We simply didn't notice at first, until some months elapsed and there became something of a pattern with deals falling through or at least being renegotiated. Then it clicked" he admits.

The procedure until then had, of course, been for mortgage offers to be based on valuations conducted in line with RICS guidelines. The guidelines advised valuers to seek "comparables" — similar homes sold nearby in recent weeks — and included references to a premium on new homes because of their supposedly prime condition and modern features.

The guidelines were later amended on 1 September 2008 when RICS and CML issued advice to valuers on how to quantify "incentives" like furniture packages and stamp duty payments offered by increasingly desperate developers trying to clear their stock. However, some individual lenders blatantly disregarded this new advice and issued instructions to down-value all new homes.

"We're doing this to protect our borrowers and ourselves", a Nationwide spokeswoman said at the time. "We find it hard to tell what incentives are worth, so we've issued a wider instruction."

But RICS dismissed that as unfair, as not all incentives were of equal value and some new homes have no incentives anyway. "The system exists for valuations to take regard of incentives on a property-by-property basis. But some worried lenders are being ultra-cautious. Lenders are just safeguarding themselves" says Luay Al-Khatib of RICS' valuation department.

CML says "a minority" of lenders introduced their own rules over valuations during 2008, and says it can neither control nor monitor what they do.

"Frankly I think most of the lenders are doing this, and have been for some time. If one of them tells you it isn't doing so, I suspect it may be being a little economical with the truth", says Ray Boulger of Charcol, one of the UK's top mortgage brokers.

Developers, of course, were up in arms at what they saw as a deliberate bid by the Nationwide and most other lenders to deter sales. The down-valuing was perceived as particularly harsh on specialist builders who made high-specification homes that had few or no incentives even in poor markets such as that prevailed back in 2008.

"This is another nail in the market coffin. It drives prices down and leads to a spiral of decline. Specialist builders produce innovative homes. They have distinctive features and command a premium, which could be as high as 10%. These valuations punish good quality developers who are innovative and rely on the premium for financial viability" warns Tony Dowse, Chairman of niche developer, Environ Communities.

But the suspension of "normal" valuation techniques had its roots in the over-supply of apartments, and this was another signal missed by the property industry.

As we have seen, in 2006 there were vast over-supplies of new flats in some regenerated city centres, notably Leeds. There were also allegations that some valuers colluded with builders to artificially inflate values.

Some believe Nationwide's blanket down-valuing policy fuelled the downward trend in its own monthly house price index, a key indicator at all times and seen by some as being biblically reliable during the most panic-strewn months of 2008. It is also regarded by some commentators as a key influence on the interest rate decisions of the Bank of England's Monetary Policy Committee.

The index is compiled from all mortgage offers made by the Society during the preceding month. If new-build homes are sold at prices based on artificially "low" mortgage offers, the deals feed into the index process. In theory this would be forcing averages down even further than any "natural" downward market conditions would justify.

So in September 2008, for example, figures from the Nationwide suggested UK average house prices were down 12.4% over the past year, while the Land Registry — which covers all house and flat sales, albeit only across England and Wales — showed the fall as only 4.6%.

"We need to take into account the dynamics of the current market. Prices are going down and it is resale value that we consider most important. New build homes only account for 8% of transactions over a year so it's a small proportion of the overall index. Valuation isn't an exact science anyway, and it is particularly difficult in the current market" says the Nationwide's Fionnualla Earley.

Sceptics were unconvinced.

"The market should decide the value of a house or flat, not a rule from a lender. No lender is doing this on second hand homes where they accept the market decides. So why are some doing it to new homes? The result is that getting a mortgage on a new home is becoming a complete lottery", claims mortgage broker Lee Holland.

King Sturge's valuation department put a little more light on it, and perhaps showed how the most upright valuers should behave. A report in the firm's quarterly journal *On the Market* stated:

> Valuation is not a science and will always involve elements of opinion. Valuers should be increasingly alert to the pressures of meeting borrowers' or banks' expectations; they must undertake the valuation independently, transparently and arrive at a true opinion of market value based upon the evidence available and the nature of the transaction or proposal.

Whether squeezed unfairly or not, the flurry of controversial valuation approaches adopted from 2006 should, in retrospect, have told us all that things were not right.

Missed signal 5: property scams

In this instance we as an industry did not miss the signs. We knew these scams went on, but naively turned a blind eye or at best hoped the phenomenon was a relatively isolated problem. It was not, and it infected the city centre investment market in Leeds in particular, and to a lesser extent in Liverpool, Manchester and parts of London.

The scams prospered particularly from 2003 to 2006 and most worked like this.

- An investment firm would be set up to buy new city centre apartments in bulk, usually off-plan and often at a discount of 20% to 40% on the advertised price.
- The written contract between developer and investment firm showed only a small discount — much smaller than the one negotiated — to allow the investment firm to borrow much more from a lender.
- After the money was borrowed and handed over to the developer, the agreed discount was then handed back, often in pure cash form, by the developer.
- The investment firm then sold the flats to small-scale investors or individual buy-to-let landlords who had the credit rating to buy at something approaching the full advertised price — which of course would be well above what the investment firm paid, and would be comparable with "direct" sale prices from the developer.

This sounds sharp practice but it is not necessarily illegal until you bring the role of the valuer into the equation. Notionally, as we have seen, the valuer should seek comparable properties by which to judge the value of any one particular unit; if the only comparable properties are those built in the same scheme by the same developer, there is little "independent" information for the valuer to go on.

In addition, some valuers in the period from 2001 to 2006 realised that there would be a steady stream of business if they valued in line with the expectations of the developer or investment firm selling the properties.

This led to the suspicion by mortgage lenders that they were being fed inadequate, or perhaps even inaccurate, information from valuers, whether that was intentional or not. This suspicion directly led to the so-called "under-valuation" policies I mentioned earlier, and which were applied in such a blanket way as to prejudice genuinely innovative developers, as well as those who were party to scams.

In other words, the antics of a minority of city centre flat builders and investment firms infected the industry and helped destroy the market for everyone else. Precisely how many estate agents and developers were involved will come out in 2009 and 2010 as court cases connected to some of these scams throw light on the matter.

But it is foolish to believe much of the property industry in those northern city centres in particular did not know of this, years before the 2008 slowdown, and did nothing.

Missed signal 6: the clever deals

In May 2007 there were two deals across property sectors which, with the benefit of 20:20 hindsight that we now have, should have told us something was amiss.

First, Land Securities publicly warned that the UK commercial property market had peaked and there were signs of a slowdown.

The property group, which earlier that year converted its £14.8bn UK portfolio into a Real Estate Investment Trust to safeguard its future, said the value of its investment portfolio had slowed in the second half of 2006–07 with some "lesser quality" retail assets seeing a downturn in values. Chief Executive Francis Salway said five little words: "This is a big deal."

He continued to warn us not only of a property slowdown but of a wider recession too. "Demand for London offices is still strong, but has been less so in recent months" said Mr Salway. "In retail, the market has seen a return to more normal conditions. Now that yield shift has slowed, rental values and minimising void levels are again becoming important detriments of future performance."

A few days later a second deal, this time a bigger one, was revealed to the public.

Foxtons, until that point, had been famous as an estate agency in London and Surrey for well over two decades. It was regarded as at best ruthless and at worst — as shown on a BBC investigative documentary — as a firm that would doctor landlords' agreements, lie to customers and artificially inflate property values. Now it was famous as an agency that was being sold by a businessman who was universally regarded as one of the toughest and shrewdest in British property.

Jon Hunt — who in 2005 moved into the exclusive Kensington Palace Gardens and, for a while at least, called himself Jonathan Hunt — pocketed an estimated £370m when he sold the firm in May 2007 to private equity group BC Partners, which had beaten rival private equity groups TA Associates and 3i. The selling price was said to be around £390m and Hunt owned 97% of the business.

It was considered a good price for the firm (as if Hunt might have ever accepted a bad price) because although Foxtons' story was an epic rags-to-riches tale of an enterprise that began life in a converted Italian restaurant in London's Notting Hill, it actually generated annual revenues of only around £100m, even at the peak of the housing market.

But Hunt was not a superman.

After all, when he sold Foxtons in London he concentrated on the firm's New York operation, which remained in his control. In little over a year it filed for Chapter 11 bankruptcy, showing that even its aggressive low commission approach could not beat the slowdown in the US housing market. However, Hunt was certainly sharp and more than one newspaper suggested at the time that his sale of the London operation was the sign of a housing market slowdown to come.

If only the property industry had listened and acted, especially as on the same day as the Hunt sale story broke, property website *Rightmove* revealed that house sellers across the country had produced the lowest rise in asking prices in over a year, while CML won euphemism-of-the-decade award for its prediction that there would be a "modest slowing in [lending] activity" during the coming months.

Missed signal 7: affordable housing was not working

There had been suspicions for years that affordable housing — that amorphous "cheap" housing for rent, direct purchase or shared-ownership — was not actually attracting the target audience of those with the lowest income and/or the greatest housing need.

Research by Genesishomes, a housing association, shows the problem. It analysed 2,500 applicants for its properties in 2008; all were entitled to apply, so met the criteria laid down by the Housing Corporation, then the body presiding over associations' tenancy and purchase processes.

Amazingly, almost half of applicants looking for homes had a household income of between £20,000 and £30,000 per year, and a further 11% were in the £40,000 to £60,000 income bracket. More than 24 different job sectors were represented in the 2,500 applications, with 20.4% working in a management or administration, 8.2% in finance and 6.8% in marketing, media and publishing — under 20% worked in the emergency services, education and health services combined, which many members of the public regarded as the core clientele for affordable housing.

"The market for affordable housing has changed enormously over the past five years with more and more people finding themselves unable to buy a home outright. The result is a wider spectrum of applicants from all walks of life. Of course, we still have a strong core of key workers applications but many other people can qualify. We have applications from accountants, architects, engineers, journalists, pharmacists and TV producers. We've even had a reptile breeder apply", says Genesishomes' Deputy Director Sharon Cummings.

Genesishomes is one of the few associations to reveal figures and is rare in expressing such aggressively market-oriented sentiments, but its profile of would-be purchasers is regarded as broadly typical of many other registered social landlords (RSLs). Indeed, the new HomeBuy scheme — the latest in a raft of government initiatives to assist first time buyers — allows associations to help anyone with an income of £60,000 or below so long as they are considered to be priced out of buying in the local open market.

Until 2007 most affordable shared ownership schemes made overt attempts to target key workers, sometimes specifying eligible jobs — typically in the emergency services and teaching — and even stating geographical boundaries specifying where the jobs and homes to be purchased had to be located, as well as income maxima. Now those criteria have been relaxed to include workers on much higher salaries and working in the private sector but still earning too little to buy on the open market.

"This has to be looked at. In many areas households earning £50,000 per year still cannot afford to buy and they do have genuine problems. But by allowing these to apply for intermediate [ie shared ownership] housing means there's automatically less chance of those earning £25,000 or less from getting on the housing ladder. There must be new initiatives to appeal to that less well off group" says John Tennor, a housing consultant specialising in affordable homes strategies in European housing markets.

According to some industry analysts, the problem in the UK is caused by the inappropriate stock held by RSLs, limited government funding for them, and Whitehall rules for shared ownership schemes failing to target the lowest-paid.

"Over the last decade it's become increasingly hard to provide two-bedroom or above intermediate housing that is affordable. The predominant supply ... has been for one-bedroom units. Grant rates need to be substantially raised to enable housing associations to procure family sized intermediate units" explains Peter Emmerton of Cushman & Wakefield, which is working on other affordable shared ownership models for RSLs.

Sue Cocking, who heads Knight Frank's affordable housing team, says there should be greater flexibility in RSL funding to target applicants in greatest need.

Currently, many RSLs insist on minimum 25% or 50% purchase options, which in high-priced areas may preclude less affluent would-be purchasers. "It would be useful to develop rented and ownership products where the rent or equity stake could vary quite dramatically according to different household circumstances. Perhaps we could move to a system where grant follows the household in need, rather than the product", she suggests.

Of course, a further option would be to build more homes so long-term pricing becomes stable and in line with incomes. "In countries where there is no housing shortage, affordability is nowhere near as much of a problem" claims Cocking.

That situation seems further away than ever, because affordable stock shortages are now worsening as a result of the housing market slowdown.

As developers have downed tools on private projects, so work on allied section 106 schemes has stopped too. In many cases, section 106 deals between developers and local planners are being renegotiated to reduce the affordable elements of schemes once the market picks up and building resumes on sites that have been significantly down-valued in the interim.

Of course there remains the social rented sector for those unable or unwilling to buy, or who miss out on shared ownership schemes where demand outweighs supply. Many discounted schemes offer rents 20% below market rates, for homes that are often larger and arguably built to a higher standard than private rented units. Figures are scant but this is now the main "key worker housing" provision in many areas.

Meanwhile, the income criteria for intermediate, affordable shared ownership is growing ever-larger. London Mayor Boris Johnson says eligible households now include those earning up to £72,000 a year, allowing still larger numbers to apply while — in the downturn at least — the number of new shared ownership homes being built remains limited.

Where, you might ask, were the genuinely poor or the key workers, at whom these schemes were originally aimed? If only we had considered the implications of even "affordable" housing becoming unaffordable to many outside the professional sector, we may have been better warned of the pain to come as prices inevitably fell.

Missed signal 8: the pain in Spain

A visit to Spain has been a miserable experience for property professionals for the past few years — but we just didn't notice the problems at first, and then we failed to interpret what we saw as warning signals for the UK.

Budget airlines now say they have plenty of spare seats to Spain because there are fewer second home owners travelling to their properties, and almost no prospective new buyers. Ryanair flights to Valencia (at one time mooted as a new holiday home haven) have been suspended, and the airline gives the slump in second home purchases by Britons as a key reason.

In airport arrival halls, the estate agents in beach shirts and shorts who used to harass visitors have disappeared — relief for those waiting at luggage carousels who used to be harangued, but a sign of a drastically smaller property industry.

Driving down the N-340, the road through the Costa del Sol, you pass terraces of new homes packed onto hillsides. Many are completed but empty; most others are part built but abandoned.

Walking through tourist havens like Fuengirola and Torremolinos you spot estate agents' offices, empty and grubby. Many have shut for good; optimistic ones have signs saying they will re-open when — some say "if" — the market improves.

Like everywhere else, this nation has suffered from the global credit crunch but what should have triggered off alarm bells in the UK were the longer-term problems of poor bank liquidity, over-supply of homes in selected areas, and disillusioned buyers.

Spanish banks' exposure to house building is higher than in the UK, ranging from 25% to 50% of their balance sheets. Some 18% of the gross domestic product is tied up in construction and associated industries. But the real shocker is the number of new homes; over three million built between early 2004 and the start of 2007, half of them on the coast. Some 800,000 were started specifically in 2007, despite the absence of any obvious demand from international buyers already tiring of Spain's vicarious market.

Those who bought after 2003 and wanted to sell quickly have accepted price falls of up to 40% say local estate agents. Developers like Victor Sagué of Taylor Woodrow conceded that, in his words, people "have had their fingers burnt". That may just be the understatement of the recession.

The firm — which builds homes in the downturn markets of the

UK and US as well as Spain — had to cut prices on 70% of its Spanish stock by as much as 25%. However, most estate agents say anyone brave enough to buy a new home should be able to negotiate further big reductions whatever the "official" price fall has been.

As if the new-build over-supply was not enough, there are also those homes on sale which have been repossessed and being auctioned by banks.

Two-bed flats with balconies typically cost £77,000 or less says the sales website *www.propertyinspain.net*. "Many are just a few years old on developments with communal pools and grounds, secure parking and close to beaches" says website spokesman Kevin Barnett. He says banks "want only their loan and repossession legal costs back."

It is therefore unsurprising that most sellers cannot shift their homes. They are reluctant to drop prices because they want to clear their mortgages, but without big falls they simply cannot compete with cheap repossessed homes on the market at bargain basement prices.

Briton Richard Netherside put Calquico, the large restored farmhouse he lives in just north of Barcelona, on the market for €945,000 in August 2007. It is a handsome property, with three storeys of accommodation, mountain views, a large outdoor pool and big gardens, too. But it was bad timing for a sale, coming just as the Spanish market slumped badly, a few weeks before the Northern Rock crisis in the UK, and while Sterling's exchange rate against the Euro deteriorated.

He registered with 10 nearby estate agents (Spanish sales are frequently conducted through multiple agents) but they were so pessimistic about his sale prospects that only two bothered visiting. By May 2008 he had no viewings. The price was then cut to €830,000 and then to €749,000, before dropping further. "With the UK market falling away and the local buyers unable to get loans we accept we are probably in for an extended wait yet" says Netherside.

Kevin Sheehan, a Surrey businessman, has been luckier, although he had to work hard to sell his five bedroom holiday villa in the Valencia region.

It was originally advertised in the summer of 2006 for €420,000 but after a year there had been "nothing even remotely looking like an interested buyer". Some potential buyers were deterred by well publicised problems for Britons whose homes in the Valencian region had been compulsorily purchased. "That doesn't apply in the area near my home, but of course it's another reason why people won't buy", he admits.

In August 2007, Mr Sheehan took matters into his own hands and spent the month contacting agents across the world in a bid to attract any interest. He sent 32,000 e-mails — yes, 32,000 — registering the villa with agents and on free sale-by-owner websites.

"I had a large number of speculative offers of half the price but I held my nerve. I reduced it down to €320,000 but wouldn't drop further. In the end I sold it to a man from Jersey for fairly close to my asking price", he says.

Experts suggest these are just two typical examples amongst tens of thousands.

"There's going to be carnage in the market. People have stretched themselves too much and developers have built one ugly scheme after another. Large parts of the coast are just wrecked" explains Mark Stucklin, a British property analyst who lives in Barcelona and runs a website for buyers, *www.spanishpropertyinsight.com*.

"On the Costa del Sol, supply has just exploded and most homes have been built on borrowed money. The developers have mortgages they'll never, ever repay. They owe far more than they'll get from buyers given the large scale price falls" he says.

On top of it all, the exchange rate has worked against sellers in Spain hoping to find new buyers from Britain. In October 2007 a €150,000 apartment would have cost about £104,000; by late 2008 it was £138,000 thanks to the strengthening Euro.

As a result of the crash, many areas with new homes look like ghost towns and this is not confined to the costas.

A new suburb of the inland Catalan town of Vic, midway between Barcelona and Girona, lies empty. It should by now have hundreds of new homes for Spanish professional owner-occupiers but acres of Vic's building sites are desolate, the show homes are closed and fencing erected. Only the alsatians remain to guard the place.

As in the UK, one of the biggest sectors to suffer has been estate agency.

The worst year was 2007. Of 80,000 that operated in January that year, only about 40,000 estate agencies survived until the following Christmas, leading to the loss of 100,000 jobs, according to the Superior Council of Real Estate Agents, a quasi-official industry body.

Many of the agencies that collapsed were small, sometimes just a person with a mobile phone, and emerged solely to cash in on the Spanish construction boom since 2002, said the Council's President, Santiago Baena.

"The ones that closed are the upstarts, the ones who came into this sector because they saw easy money", he said. Larger, well-established companies have closed some branches but are still at least operation.

Sad? Of course. Predictable? Yes, at least in hindsight.

Now we know all about Spain, over 40,000 workers lost their jobs each week in late 2008, official unemployment at 13% of the workforce or now past the magic three million mark. But what is perhaps most shocking to remember is that many of these problems were happening as long ago as 2005.

Did we not notice? Did we not care? Could we not see that, despite the differences between the Spanish and British markets, there was enough in common to make us feel worried?

So there were signs, in the UK and overseas. Other commentators may include other signals that were equally obvious in retrospect, and equally these were missed.

Yet, even those people who began to worry early on, ahead of the gloom that spread like treacle over the industry from the spring of 2008, admit that the scale and speed of the problem took them by surprise.

"There were some — myself included, I think — who felt they did see signs that there was trouble ahead. But I know of no one who expected the double whammy of the housing bubble bursting and the banks turning off the lending tap" says Henry Pryor, a housing market analyst who began his property career as an estate agent at Savills and Strutt & Parker, but who now runs a series of real estate websites.

"These factors happening at the same time were enough to stop the market in its tracks, and transaction volumes went down by as much as 70% from 2006 levels. We went from around 14,000 mortgage products in August 2007 when Northern Rock hit the rocks to 5,000 by Christmas [2007]. There are now under 3,500", he says:

> Today's problems are not that houses are unaffordable as such but that those who want to buy cannot borrow despite their circumstances being no different than they were two years ago. Many said things were different this time because interest rates were low rather than at 15% as they were when the previous bubble burst. This is simple supply and demand. Demand has dried up because it can't borrow, and left an overhang of supply which is driving down values — and now rents too.

The UK property crisis: from downturn to recession

We were not to know about the full onslaught about to hit us when, in the spring of 2007, Jon Hunt was selling Foxtons, lenders were changing attitudes to valuations, and flats were being built in large numbers in many of our city centres. From that point onwards, however, things began to happen — and fast.

Whole areas of international finance that had passed us by before started appearing near the top of news agendas.

After a two-year period between 2004 and 2006, when US interest rates rose from 1% to 5.35%, the US housing market began to suffer and in the summer of 2007 prices were tumbling by as much as 40% in some large cities with high jobless levels. There were also wild rises in the numbers of homeowners defaulting on their mortgages.

As a result, the financial pages, sometimes even the front pages, began to talk about default rates on sub-prime loans — high risk loans to clients with poor or no credit histories. Some commentators even began to say that these deals had themselves been sold on by the lenders to other institutions.

Suddenly, an obscure financial problem across the pond was a real threat to our everyday lives. Then everything became something of a blur.

August 2007

The crisis

The European Central Bank poured in funds to the French banking system to neutralise what BNP Paribas called "a complete evaporation of liquidity"; UK sub-prime lenders started to halt lending to new sub-prime applicants and raised mortgage costs for existing ones; German bank Sachsen Landesbank, a major sub-prime lender, was taken over by a rival after declaring itself near collapse.

UK property

A then-typical monthly total of 109,585 homes were sold; the Land Registry said English and Welsh house prices were rising by 9.4% per year; at least 60 new estate agency branches opened; the Bank of England base rate was at 5.75% and a typical variable rate mortgage was 7.75%; the Nationwide Building Society's house price index valued an average UK home at £183,898; Gordon Brown announced targets of three million new homes to be built by 2020, even though only 122,000 homes had been built so far in 2007 — a 2% rise on the previous year.

September 2007

The crisis

The run on Northern Rock saw £1bn in savings withdrawn in one day, creating queues outside branches and attracting global attention; the Libor rate (the interest rate charged between banks which lend to each other) rose to 6.7975%, the highest for a decade, as institutions became more suspicious of each other's financial viability; another German bank, IKB, announced a £550m loss because of sub-prime mortgages; the US base rate dropped to 4.75%; and the Bank of England injected £10bn to stabilise UK markets.

UK property

The Land Registry reported prices still rising 8.7% annually and the Nationwide house price index took a typical UK home's value to £184,723; sales volumes dropped by over 22,500 to 87,013; the Bank of England base rate was still 5.75%; estate agent Savills said its country house market was starting to grind to a halt, while the developer Bovis admitted that new-build house price growth was now confined to London, Scotland and Northern Ireland.

October 2007

The crisis

The month started with UBS announcing £1.7bn losses from sub-prime defaulting, the highest profile bank to date to admit sub-prime exposure; then Citygroup admitted a £1.5bn loss before writing down a further £3bn; Merrill Lynch announced a £4bn exposure to "bad risk debts"; the International Monetary Fund warned that the UK average mortgage loan-to-value was 76%, not much below the US's 84%.

UK property

House prices in Northern Ireland had risen an eye-watering 54% in a single year, reported the Nationwide, which said a typical UK home was now priced at £186,044; the Land Registry said annual house inflation had dipped slightly to 8.1%; sales volumes for homes were still low at 90,281; the US arm of hackle-raising London estate agent Foxtons filed for bankruptcy in Manhattan; back home, upper-end agency Cluttons said the London property market had peaked.

November 2007

The crisis

Citigroup, which was to become a high-profile casualty of the crunch more than a year later, admitted billions of dollars of losses from sub-prime loans and Chuck Prince, its Chairman and Chief Executive, stood down; in Britain, the Bank of England announced its mortgage approvals were at a three-year low.

UK property

The Land Registry said annual house price growth was static at 8.1%, but monthly sales volumes were still below average at 90,581; the Nationwide Building Society ordered its panel of valuers to "down value" new flats to second-hand levels, and its price index showed the first fall as the average UK home dipped to £184,099, a £2,000 drop; CML warned that lenders may have to sharply reduce the number of mortgage products; by contrast, the property sales website *Rightmove* announced that asking prices rose 2.7% over the last month; "Agents are pushing a little too far" warned veteran London estate agent Simon Albertini.

December 2007

The crisis

Credit-rating agency Standard & Poor's downgraded its investment ratings for many bond insurers, effectively weakening the "guarantees" that existed to repay loans if the insured bond-issuers went bust; the US Federal Reserve co-ordinated attempts by five central banks to lend funds to individual banks; the Libor rate dropped sharply in UK early in the month, bouncing back up by Christmas.

UK property

The year ended with another price fall according to Nationwide — down another £2,000 to £182,080; the Land Registry still showed a full-year rise of 6.7% but it, too, said prices in the month of December itself dipped slightly; only 72,880 homes were sold in the month and *Rightmove* claimed homes were now taking an average of 85 days to find a buyer, the longest period since 2002.

January 2008

The crisis

The US Federal Reserve cut 0.75% off interest rates — the largest reduction for 25 years; on 21 January, global stock markets saw their largest falls since 9/11; Scottish Equitable warned commercial property investors that they may have to wait a year to withdraw funds; bond insurer MBIA announced a $2.3bn loss.

UK property

The Land Registry showed a monthly price rise and average homes still cost 6.4% more than a year earlier; Nationwide reported a minor monthly price fall, with the average home down to £180,473; there was yet another big fall in sales to 53,221 — scarcely two thirds of the January 2007 level.

February 2008

The crisis

The G7 said worldwide banking and sub-prime mortgage losses could be $400bn (in retrospect, if only); the US Federal Reserve made $200bn available to banks to improve liquidity; Northern Rock (or at least its losses) were privatised and 10% of the UK's mortgage products were withdrawn in one month.

UK property

The Nationwide average price fell again to £179,358 and the Society revised its 2008 prediction from no change to minor falls; CML said repossessions in 2007 were the highest since 1999; transaction levels were up slightly to 57,016; the Bank of England base rate was cut 0.25% to 5.25%.

March 2008

The crisis

Now things got serious. On Wall Street, Bear Stearns was bought by JP Morgan Chase for $240m, one eighth of which was itself a loan — a year earlier, Bear Stearns had been worth $18bn; Lehman Brothers' stock slumped 11% in one day thanks to uncertainty about its credit exposure; President George W Bush described the US economy as "sound".

UK property

Sales were down to 53,080, the lowest for a decade; the Land Registry price index showed only a 3.6% increase on the year, while a typical home dropped to £179,110.

April 2008

The crisis

Another fifth of UK mortgage products were withdrawn in four weeks, including the final 100% mortgage; CML predicted mortgage lending would be halved in 2008; the Bank of England announced a £50bn scheme to allow banks to swap mortgage debts for secure government bonds; the UK Government announced a 54% rise in firms going into administration across the economy; the International Monetary Fund said worldwide credit crunch losses could exceed $1 trillion; the Bank of England base rate fell again, to 5%.

UK property

Persimmon announced a 25% fall in sales since 1 January, forcing major job cuts and profit losses, hot on the heels of 600 jobs being lost at Taylor

Wimpey and almost 400 at Bellway; RICS said members' confidence in the housing market had hit a 30-year low; the Land Registry reported house prices still rising 2.4% annually but the Nationwide index showed its first monthly fall since 1996, with an average home now costing £178,555; Savills became the first agent to "tell it as it is" when it warned house prices may fall 25% in the long-term.

May 2008

The crisis

On 1 May, the Bank of England Deputy Governor Sir John Gieve said: "The most likely path ahead is that confidence and risk appetite will return gradually in the coming months", indicating the worst of the credit crunch was over — but on 18 May Jean Claude Trichet, President of the European Central Bank, claimed the worst was yet to come with an "ongoing, very significant" correction comparable to the oil crisis of the 1970s; Swiss bank UBS launched a $15.5bn rights issue to try to cover 50% of its US-derived sub-prime losses.

UK property

Estate agent Knight Frank said that come September, prices of some London homes may show rises again.

June 2008

The crisis

Barclays announced £4.5bn share issue; IndyMac collapsed, becoming the second biggest bank failure in US history; UK holiday and car industries issued first figures suggesting lower-than-expected demand.

UK property

Sales volumes only just exceeded 54,000; residential rents rose across the UK by an average of 5.7% in a year, according to a survey by website *Findaproperty*, but consultancy Hometrack said the rise was 10%-plus in big cities.

July 2008

The crisis

CML revealed that the UK mortgage market now had only 50% of the products it had available a year earlier; there were the first signs of US Government assistance to Freddie Mac and Fannie Mae, which together were guarantors of $5 bn of US home loans; only 8% of HBOS investors bought into its £4bn rights issue — not surprising, as it was priced higher than the bank's existing shares; the FTSE fell 20% over the month.

UK property

The Nationwide extended its "down-valuing" policy on new homes from flats to houses; the Land Registry index turned red as prices in England and Wales were reported 2% lower than a year earlier; Nationwide's index showed prices 8.1% lower than in July 2007, with a typical home at £169,316.

August 2008

The crisis

HSBC, which had substantial exposure to the UK and US housing markets, reported a 28% fall in profits and warned that global financial markets were constipated; the UK gross domestic product remained static; Bradford & Bingley posted losses of £26.7m for the first half of 2008; and the UK Government officially admitted that the downturn would be "profound and long-lasting".

UK property

House prices fell 2.7% and first-time buyer numbers dropped to an all-time low according to the Department for Communities and Local Government (DCLG). It said flats dropped 5.1% that month; the Halifax's price data showed another 1.7% off during the month and average values were down 12.8% on the year.

September 2008

The crisis

On 29 September, another 12% of the UK's remaining residential mortgage products were culled in eight hours — at 9am there were 3,252 (itself a fraction of the total a year earlier) down to 2,988 by 5pm. Buy-to-let products fell from 622 to 481 in the same period; in the UK, Lloyds was taken over by HBOS, subject to detailed negotiations, and Bradford & Bingley was nationalised; in the US, Lehman Brothers filed for bankruptcy, Merrill Lynch was taken over by the Bank of America, and Wachovia was bought by Citigroup.

UK property

Kier, the UK's 15th largest house builder, ended construction of private homes to concentrate on social housing and regeneration schemes; RICS set out a 10-point plan to "save the housing market"; a £250m redevelopment of Brighton Marina, including 853 flats plus retail and entertainment space, was mothballed; some estate agents admitted that they advised sellers to underwrite values of properties to persuade buyers to commit, effectively compensating purchasers for future losses; the *Financial Times* announced that private equity firm BC Partners, which acquired Foxtons for £390m in May 2007, had expected a worst case sales drop of 30% — but the actual drop was over 50%.

October 2008

The crisis

In retrospect, this was the pivotal month for the credit crunch — leading to worse in later months for the UK property industry. The US House of Representatives passed a £394bn plan to rescue American banks; German, Danish, Israeli, Peruvian, Icelandic, Swedish, Swiss, French, Indian, South Korean and other central banks followed suit; in the UK, the Financial Services Authority raised the limit of the amount of deposits that were guaranteed should a bank go bust to £50,000; the number of people out of work in the UK soared to a 17-year high; data analyst firm Begbies Traynor put 323 British retailers on its "critical watch list" which it defined as having a 70%-plus chance of failing within a year.

UK property

Her Majesty's Revenue and Customs said there were just 59,000 property sales in the previous month, down 53% on the year — even after the Government raised the stamp duty threshold; Barratt Developments announced 43% discounts on five or more flats at certain schemes; Dresdner Kleinwort consultancy said 40% to 50% reductions in city centre flat prices were now common; owners on a new private estate in the Cotswolds complained that developer Bovis Homes was selling unsold houses to a local housing association — other developers followed suit, although many associations rejected what some called "sub-standard flats"; Cluttons announced a new contract for second-hand home buyers to "prevent" gazundering; Imagine Homes, the buy-to-let investment firm run by Grant "Mr Anthea Turner" Bovey, was taken over by HBoS after two years of multi-million pound losses; Caroline Flint was sacked as Housing Minister and replaced by Margaret Beckett.

November 2008

The crisis

The Bank of England base rate was slashed from 4.5% to 3%, the largest interest cut since 1981; Barack Obama was elected US President; the Bank of England slashed interest rates from 4.5% to 3%, taking them to a 50-year low and beating even the European Central Bank rate, which was lowered to 3.25%; China and Pakistan joined the party by announcing "stimulus packages"; the Eurozone officially entered recession; the US announced a £13.5bn rescue deal for banking giant Citigroup; the French Government agreed to buy up to 30,000 new private homes unsold by developers; India revealed price falls of 30% to 50% on new-build homes.

UK property

Taylor Wimpey revealed that its total 2008 job losses were 1,900, not its previously stated 900; Savills reiterated that it may take a decade for some regional prices to return to peak late-2007 levels; Nationwide's monthly index showed prices falling 0.4% in November, much less of a fall than October's 1.3% — but was a false dawn.

December 2008

The crisis

UK VAT came down to 15% as part of the Government's "fiscal stimulus"; the Bank of England dropped the UK base rate to 2%, the lowest level since 1951; Dubai developers DAMAC and Nakheel laid off workers; Credit Suisse laid off another 5,000 financial sector workers worldwide; President Bush said the US Government would use up to $17.4bn of the $700bn meant for the banking sector to help General Motors, Ford and Chrysler; Bank of America announced 35,000 job losses; back in the UK, the retail chains of Woolworth, Whittard, Zavvi, MFI and clothing stores Adams, Officers Club and USC all confirmed that they had gone into administration.

UK property

Sandersons of Darlington, one of the UK's oldest agencies, shut down its sales operation; Chesterton was bought outright by Mercantile Group, and merged with Humberts; the Halifax house price index dropped 2.6% in a month taking an average home to £163,605 — the same level as in July 2005; eight banks backed Gordon Brown's scheme to cover interest payments on mortgages of home owners temporarily unable to pay because of the downturn; the Government announced that £400m was now available for first-time buyers through HomeBuy Direct, up from £300m announced in September; Hometrack, after decrying rival indices for exaggerating price drops, fell in line and predicted a 2009 10% drop; Halifax and Nationwide refused to predict the UK housing market in 2009; the British Bankers' Association confirmed a full-year 60% fall in mortgage approvals.

January 2009

The crisis

Japan's Government predicted that its economy will have zero growth in the year ending March 2010; French statistics agency Insee said that France's economy shrank 0.8% in the fourth quarter of 2008, and will contract another 0.4% in the first quarter of 2009, taking it into recession; the World Bank projected China's growth to slow from 11.9% in 2007 to 7.5% in 2009, while India's growth prospects will be cut from

9% to 5.8%, and most of the world — including smaller emerging economies — risk recession.

UK property

A survey showed a large minority of homes on sale for over a year — 26% in Rochdale, 23% in Aberystwyth, 20% in Swanage, 18% in Pontefract, showing the problem is UK-wide; Land Registry data showed that English and Welsh house prices fell 12.2% in the year to November 2008 — the East Midlands and East Anglia saw a 14% drop, London 10.2%. Figures compiled by Estate Agency News show that in the past year, the 10 largest UK agents closed 399 branches between them. Halifax closed 84, LSL lost 82, Countrywide shut 72 and Spicerhaart closed 50.

24 January 2009

After two quarters of negative economic growth, the UK formally entered recession.

2 When Estate Agency Lost its Mojo

Here is another date, this time not a particular landmark of the credit crunch.

Friday 7 November 2008, indeed, was just another typical day for Robert Billson, the estate agent who runs the Savills office in Nottingham — or at least it was just another day in the un-typical great housing market slowdown.

"I went to five sellers in a row telling them their properties were on sale at too high a price. One of them had originally been on almost a year ago at £595,000 but the sellers had rejected an early offer of £575,000 and then one of £550,000 and then one of £525,000. Then it was on at £495,000 and they rejected an offer of £475,000. Now they just want rid of it. It's no fun being an estate agent when the days are like this", he admitted.

Billson, a well known realist in the residential agency, had spent much of the year telling sellers the same story, in line with Savills' well documented "tell it as it is" approach to the slowdown. He and the firm survived the crisis year, as it would do given that only 20% of its income is derived from the residential sales process.

But while few would envy Billson's task at telling vendors that things were bad and getting worse, he at least still had a branch office. Plenty of others elsewhere did not.

Certainly, there had been no shortage of predictions of how many would go out of business by the end of the slowdown, and the roll call of office closures suggests that many of these predictions have been met and perhaps exceeded.

For example, the Centre for Economics and Business Research (CEBR) forecast early in the slowdown that some 15,000 estate agents across the UK would lose their jobs within the first full year of the credit crunch. If you added in ancillary professional staff — surveyors, administrative staff, show-suite people, marketing types — the total would exceed 40,000. Just before Christmas 2008, the CEBR pushed that total to 50,000.

Jorg Radeke, one of the authors of the original report for the CEBR, said:

> Although unlikely to be the victims of the credit crunch that will garner the most sympathy, estate agents and others involved in managing real estate are likely to find the next [period] particularly tough and there will be extensive job cuts. The only silver lining for this part of the business services sector is that 'what comes down must go up' and as real estate is among the first to face the economic downturn it will also be among the first to benefit from a future economic upturn.

Debtwire, one of those anonymous business information companies collecting data on market sectors, predicted early in 2008 that 1,000 estate agencies had already gone out of business so far that year with another 150 closing up shop each month.

Even the National Association of Estate Agents (NAEA) — an organisation that, despite its title, lacks a history of significant or authoritative research on the industry — came out with a gloomy forecast early in the slowdown. It predicted that the number of estate agents in the UK would shrink by 25% by the start of 2009 with at least 6,000 individuals losing their jobs. "Once the market starts picking up after a few years people will invest again, but this cycle usually takes three to four years" said NAEA President Chris Wood.

In fact, the story was much worse than any prediction suggested.

By early 2009, the *Daily Telegraph* newspaper, using a private researcher, calculated that one in four agency offices across the UK had closed and some 32,000 estate agents had been made unemployed in the preceding 18 months. Only a small proportion of those office closures had occurred before the summer of 2008 but from that time onwards the rate accelerated sharply.

"I think the figures are reasonably conservative. It really has been a bleak year, and the losses faced by the estate agency industry have been mostly swept under the carpet", the researcher said.

As a result of the contraction of the housing market, agents (make that former agents) kept popping up doing other business, including many seeking freelance consultancy assignments. Jonathan Mullin of recruitment firm Holland & Tisdall told the recruitment industry journal *Recruiter* in September 2008:

> The number of estate agents we have placed has gone up by around 50% in the last six months.

Angela Webb, Director of Recruitment at rival company Michael Page, said:

> The people who typically fit into our organisation will be from corporate sales.

It is a sign of the unstructured, informal nature of the UK's estate agency industry — no formal qualifications required to become an agent, no licencing required to trade, and no single professional body that commands loyalty from the majority of practitioners — that no one knows precisely how many agents have gone out of business since the downturn began.

What is clearer to see is that those agents that did suffer, either by severely reducing their branch network or simply closing down completely, were in the main following a traditional business model.

This typically involved one or more high street branches, staffed by semi-skilled agents with or without formal training, on a relatively low basic pay and reliant for between 35% and 80% of their income on commission. Online activity would almost certainly be limited to publicising vendors' homes on the company's website and a handful of "umbrella" portals like *Rightmove* or *Fish4Homes*, and possibly the upmarket *Primelocation*. There probably would not be any substantial income derived outside of property sales as few of these failing businesses had lettings, property management or consultancy divisions.

We shall see in Chapter three whether that typical business model was best suited to 2008 — and certainly how it will have to change in the future — but we will first try to spot what common denominators existed amongst those agencies which appeared to fair worst during the slowdown.

Estate agents falling like autumn leaves — all year round

One of the biggest shocks on the downturn was the decision by HBOS to shut a total of 84 of its Halifax estate agency branches. This meant the loss of at least 550 posts, although some individuals — mainly mortgage advisors — ended up being transferred to the group's banks.

Shortly before the announcement, the HBOS banking division had reported a 72% collapse in half-year profits and had received a financial drubbing when it attempted to raise £4bn from its shareholders ... only 8% of which took up the offer.

The firm's official, and characteristically euphemistic, comment was that it needed to "reshape its business in the current market conditions, following a significant decline in housing transactions". It would in future be concentrating on its core markets in the Midlands and the North, where it had its remaining 151 branches, and would try to improve its market share in secondary locations where it already had a presence.

Almost as shocking as the Halifax news was the announcement by Your Move, the country's third-largest estate agency, to close 10 of its franchised branches mostly in affluent parts of the South West including Taunton in Somerset and Clifton in Bristol. Although the numbers of jobs involved were relatively few, this news came just a few months after Your Move's parent company — LSL Property Services, which also owned Reeds Rains — had closed 12 other branches and sacked 315 job losses.

Countrywide, the UK's largest estate agency group specialising in lower priced homes but including high street names dealing in all sectors of the market, like John D Wood, Bairstow Eves, Gascoigne-Pees and RA Bennett & Partners, axed 50 branches in late 2007 and early 2008 and has continually warned that the rest of its empire remains under review.

Connells' estate agency business — regarded as a successful survivor of the downturn — nonetheless barely broke even in the first half of 2008 even though, it is worth remembering, its non-sales side of the business made a healthy £10m profit. But fee income from house sales was down 44% and "our market was about half what it was this time last year", according to Chairman Stephen Shipperley.

In the year to late-2008, Connells had shut at least 20 offices and lost 850 staff overall, while Shipperley went on record to berate the parallel universe in which many of his rivals operated:

> One of the issues I have with people in this industry is that many estate agents do not believe prices have fallen significantly or do not persuade their clients to lower expectations. The falls in price vary according to location and type of property. I think city centre flats have suffered some of the steepest falls, perhaps 25%.

Spicerhaart, the third largest estate agency group, closed nine branches early on in the downturn and followed this with another 40 by the end of 2008. Througout that year, many other agents' closure announcements came thick and fast.

In London the only UK branch of German agent Engel & Volkers was shut at short notice — only two years after it boasted in a reception to journalists at The Ivy restaurant that it was going to expand across Britain, concentrate on selling new apartments and try to rival Knight Frank as a popular agency for the thrusting young professional investor.

Kinleigh Folkard & Hayward, one of the capital's biggest chains, shut its one-time flagship Knightsbridge office in Brompton Road and reduced its staffing from 720 to 480 in just nine months, while lower-end agent Felicity J Lord closed its New King's Road office. Friend & Falcke sold its office at 96 King's Road and closed its Fulham branch, while its veteran Managing Director Simon Albertini said:

> We've been operating for almost half a century through awful recessions. But none as bad as this.

In early summer of 2008, Jonathan Vandermolen, founder of Blenheim Bishop, sacked 10 staff and closed down the sales side of his much publicised Mayfair agency:

> I couldn't make it work. Income from new homes was 75% down on the previous year and my view of that market is not good long-term.

Two years earlier, Vandermolen had sold Blenheim Bishop to Humberts — of which, more later — for an estimated £2.3m but he later bought back his part of the business for a token £1.

"Most agents operate on the basis of a 25% to 30% profit margin. If trade is 50% down and 50% of your costs are for premises, staff and advertising, then it becomes difficult. If we have a couple of years of this, perhaps 50% of all estate agents will go out of business", he told *Estates Gazette* Magazine.

In the north of England, Hunters started 2008 with 300 staff in 22 offices and an annual sales rate of 1,600 homes, but by early 2009 had only 180 staff in 16 offices, after a year in which just 800 properties were transacted.

Amongst the many hundreds of office closures in rural England, three were by Dreweatt Neate — Hungerford, Wantage and Devizes — just before the firm pushed ahead with a merger with Carter Jonas. In the Cotswolds, both Hodsons and Andrews announced a mix of branch closures and job losses.

Wales' largest estate agency, Peter Alan, slashed its branch structure and staffing levels by almost 15% towards the end of 2008. Peter Griffiths, Chief Executive of the agency's parent company, the Principality, said the housing downturn was the worst the company had seen in over 20 years of trading:

> The frustrating thing for Peter Alan staff is that they are experiencing high levels of interest from the public. People want to buy, are in a financial position to buy, but can't get a mortgage.

Scotland was the same. Clyde Property severely cut back on its sales side and retrained staff to work in its lettings business, while the much larger Slater Hogg & Howison group made over 70 staff redundant. The Glasgow Solicitors Property Centre (remember, solicitors are de facto estate agents in the Scottish property system) slashed 10% of its jobs in 2008 while Edinburgh-based Stewart Saunders — a 32-year-old business with offices in the city centre, selling homes across the east of Scotland — closed for good.

Northern Ireland fared no better. Scores of agents jumped on the bandwagon of the region's extraordinary 50%-plus leaps in house prices between 2003 and 2007; many had no previous experience of property sales and almost all newcomers avoided going into other less profitable but ultimately more reliable activities, like rentals or property management. The largest branch closure programme in the province was by US-owned Remax, which in early 2007 had 18 branches in Northern Ireland, yet by Christmas 2008 was down to just eight, creating an estimated 100 job losses.

Of course, each closure or reduction in size for any agency anywhere involved pain for the individuals and families concerned, and every managing director's decision was based on slightly different circumstances submerged beneath the all-embracing title "credit crunch". But there were some common denominators.

- First, most of the agents' offices that closed had no or minimal lettings divisions. Although eventually the rental market ran into trouble as frustrated sellers put their homes on the lettings' registers, for many agents with a lettings outlet the principal source of income for 18 months or so was from rentals.
- Second, and closely related, very few of the smaller agencies had any form of property management business. This did not apply to the large chains (some individual offices that shut did have these responsibilities) but ancillary work of this kind was simply spread amongst the remaining offices, so did indeed help the company's overall income. The big losers were the small one- or two-office firms which relied solely on selling.
- Third, many of the agents that closed quickly were ones that had not invested in new technology and offered what could only be described as "old fashioned" service techniques.

Many failed businesses (not necessarily a description of all of the above) can be accused of holding on to what have become tarnished real estate conventions — ever-more high street branches, wildly over-optimistic assumptions of the market, and pile-'em-high buy-to-let investment principles. But there were also some that were, if anything, even more shocking examples of business management.

Humberts, Foxtons, *Rightmove* and Inside Track were very different types of business from each other, but were each symptomatic of the property industry's excesses since the Millennium. They performed badly during the downturn, but more important is that they were indicative of other problems — two showed a rather old fashioned approach to how they operated, while in the other two instances the problem was an aggressive (and some would say amoral) approach to the public perception of property.

Humberts

Humberts is a quintessentially British, mainly rural estate agency that has been operating for 165 years. By now its financial traumas appear to have been going on almost as long, but the short version of the Humberts saga goes like this.

In November 2005, the then 41-branch Humberts chain was taken over by the Alternative Investment Market-quoted Farley Group, which ambitiously snapped up smaller regional agencies operating in the group's core business of up-market rural and equestrian homes.

In 2006 and 2007, the then Chairman Max Ziff said he wanted Humberts to become a 200-branch empire, transforming itself from a small regional firm into a rival to the likes of Strutt & Parker in the southern half of England. Humberts then rapidly acquired some 40 branches from other agencies, some of them on a curious "deferred payment" basis. It made three major acquisitions in just one month, September 2007: Halls in the West Midlands, Thomson Currie in London, and Fox and Manwaring in Kent. All this was happening, of course, just as the housing market downturn moved into a higher gear. It could not have been worse timing.

After Humberts' shares were rising strongly in 2006 they plummeted when in 2007 there was an £18.4m write-down on the value of the newly acquired businesses, and by February 2008 the Humberts group — now 80-plus branches — reported a £17m post-tax loss and gave the first warning about its future. At the time Humberts blamed the downturn and undoubtedly this was part of the problem, although arguably causing less difficulty than its aggressive and ill-timed expansion programme.

A financial restructuring in early 2008 involved raising over £5m to ease cashflow, and a rationalisation of those new acquisitions. All this was peppered by quasi-legal acrimony. Humberts was then and remains to this day an unhappy ship and is hardly a shining demonstration that the traditional estate agency model is working.

The then new Chairman John McLean was quoted in *Property Week* in late 2008 saying:

> If you take the analogy that for three months we were a patient on the operating table haemorrhaging to death — well, we've stopped the haemorrhage. After this, we're out of intensive care but there is a significant amount of convalescing to do to get the patient up and walking. And given that the central focus ... has been sorting out the business in terms of getting our models and structures and plans in place, following this particular exercise, we can actually start to focus on running a company rather than rescuing a company.

Humberts' aggressive purchasing programme may well have led it into trouble even in the strongest of housing markets, but the company's problems were, we can now see, an early sign that the downturn was in full swing.

It came as little surprise when in the dying days of 2008 it was announced that the 34 remaining offices of Humberts were merged with Chesterton's 22 branches. Robert Bartlett, Chesterton's Chief Executive, became Chief Executive Officer of the merged company (unimaginatively called Chesterton Humberts) and his first statement in his new position promised "significant cost and revenue benefits".

Whether it will remain an agency with what one senior property consultant called "an out-dated, overly formal, uber-traditional and rather crusty" image remains to be seen. But one thing is for sure — if it is to survive, that sort of image needs to go.

Foxtons

The Foxtons UK saga is not dissimilar to Humberts in some ways. The difference is that while we can now see that Humberts' purchases in 2006 and 2007 were for far too high a figure, so here we have seen that the sale of Foxtons by the canny Jon Hunt was in retrospect for far too large a sum — good for him, bad for the new owners.

To say the legendary Foxtons brand has been controversial is to state the obvious.

One of the public relations executives hired by the firm — and coming from what has been described as one of the most expensive PR agencies in the country — allegedly admitted off the record to a journalist that his job was to cover up "mistakes and irregularities."

What other regional estate agency (and that is all it is, after all) has a national and perhaps global reputation for its sales techniques, its rough and ready staff, its high churn rate, its legendary weekly meetings where staff are lauded or lambasted depending on their success, and even its initiation ceremonies for newcomers? All this is on show in the website *www.wehatefoxtons.com*, which has become a lightning rod for dissent from within and without the firm.

So when it was sold by Jon Hunt to private equity firm BC Partners, it was thought that Foxtons may enter a period of quiet reflection, not to say downsizing. But nothing of the kind happened, at least at first, thanks to the controversial financing of the purchase.

BC allegedly borrowed £390m from the Bank of America and Mizuho, one of Japan's largest banking groups, just before both the capital markets and the housing sector hit trouble. BC's repayments have been in the order of £26m a year. By late 2008, BC remained confident it would not breach its banking covenants — a complicated ratio of turnover in relation to debt, with minimum "targets" to ensure Foxtons remained viable — but by then headcount was down by around 70 and the future of some branches looked uncertain.

More fundamental will be its debt restructuring. In early 2009, Foxtons and its brokerage arm Alexander Hall breached the borrowing agreement it had with the Bank of America and Mizuho. This means it did not meet its agreed debts-to-profits ratio and allows the lending banks the right to push Foxtons into receivership if they wish. BC Partners partner Andrew Newington was reported to have called the decision to buy Foxtons a "colossal error" and there remains uncertainty over whether the agency will have more working capital to see it through the recession.

Ironically in October 2007, before most in the UK estate agency industry had noticed even a whiff of a downturn, Foxtons' US empire was already biting the dust. At the time a legal spokesman for Foxtons North America (FNA), which set up in the suburbs of New York and New Jersey as a low commission no-frills agency in 2000, said the company was intending to oversee an "orderly liquidation of assets" and that it was unclear whether the Foxtons brand would survive in the US; it didn't.

FNA was majority owned by Jon Hunt. Its business model was based on undercutting existing estate agents by charging commissions of around 3%, barely half the traditional commission charged in that part of the US. FNA fired its 375 employees and ultimately filed for Chapter 11 bankruptcy protection. It owed just under $500,000.

Until the recession ends, there will be no guarantee that Foxtons' UK brand will not meet a similar fate.

Rightmove

Part of the theme of this book is to show, in later chapters, how some estate agents and developers have used technology to modernise their activities and shift their business model into a more robust format suitable for a post-recession Britain.

But this is not to say that companies already at the heart of that technological transformation cannot hit trouble — just look at *Rightmove*, hit by a double whammy in 2008 of reducing transactions and increasingly dissatisfied estate agency clients.

Its share price dropped dramatically in late 2007 and throughout much of 2008 as the thousands of estate agency closures led *Rightmove* to issue dramatic profit warnings. By October, the writing was on the wall, and 60 of its 300 staff were made redundant in a restructuring plan aimed at saving £5m. "The scale of the redundancies reflects the company's view that the current challenges will continue through 2009, though not at any worse level than we have previously indicated", according to *Rightmove*'s Miles Shipside.

Yet there was another sub-text. Running alongside the downturn was *Rightmove*'s own attempt to increase rates for agents advertising their (diminishing numbers of) sales properties on the website, and to make it difficult for agents on the site to suddenly withdraw. Mid-range high street estate agency branches spend up to £450 on average each month listing homes on *Rightmove* — and as belts were tightened, so the more web-savvy agents noticed new websites such as *Globrix* and *Zoomf* which would publicise homes for sale for free. They relied on on-screen and "click through" advertising revenue instead of agents' advertising subscriptions.

Outspoken veteran Trevor Kent, a warhorse with a single-office agency in Buckinghamshire and a former leader of the NAEA, wrote to *Rightmove* saying:

> Your company policy of not being fully frank with your existing members, and hiding from them the considerably upscaled rejoining cost is disingenuous in the extreme, especially when a brief holiday from Rightmove might make the difference between survival and bankruptcy. A bankrupt firm is certainly not going to rejoin when the market improves, but a survivor might. However, some will not return on principle, when they learn they are to be punished by a 59% hike in monthly subscriptions and that this rate was hidden from them when they withdrew.

Rightmove is unlikely to disappear without a fight — after all, its old business model made the people behind it very rich, with good profits reported as recently as spring 2009 — and it may yet transform its approach to agents and advertising to embrace the new "free access" model pioneered by *Globrix* and *Zoomf*.

But its structure and inflexibility has suddenly appeared very outdated in a credit crunch environment, at least when compared to new websites nibbling away at its core business. This is an ever-present danger in a conservative industry, where even people and practices which were at one time pioneering, now appear reluctant to unable to change in line with modern circumstances.

Inside Track

Few will mourn the loss of Inside Track, the company that latched on to the buy-to-let investment boom. It boasted of showing people "how [they] could give up work and be a property millionaire instead" but was an early victim of dropping British housing values — and for that matter, tumbling prices in Spain and Florida, too, where it often bought apartments off-plan in bulk to sell on to its thousands of clients.

Inside Track Seminars, which labelled itself "Britain's biggest property investment company", was set up in 2002. It specialised in holding what it called "free workshops" at hotels across the country. Lasting two to four hours each, these painted a world where anyone could become a "property millionaire". But it was a model that depended entirely on a rising housing market.

2005 was a halcyon year for Inside Track, as it made annual profits of £12.1m and claimed that over 31,500 people attended its seminars. The position may have been even rosier, as internal cross-funding between companies within the same group made transparency unsurprisingly difficult. How times changed. By early 2008 they were getting only a few dozen attendees at each seminar and internal management accounts for the nine months to 31 January 2008 — discovered by a national newspaper — showed income of just £239,000, with a £97,000 loss in January 2008 alone.

Although highly unusual, not a traditional estate agent and regarded as an investment club, Inside Track's modus operandi shows how morally bankrupt a few elements of the residential market had become in the years just before the slowdown.

Back in 2006 the author of this book received a mailing from Inside Track headed: "How you could give up work and be a property millionaire". It went on: "Start from scratch - live on easy street instead of struggling for a living!" The leaflet described how the Inside Track programme of seminars, investment teach-ins and expert guidance could enable you to retire in three to five years' time, find some of the most profitable investment opportunities — which it said were rarely advertised to the wider public — and "make money from property in good times (easy) and bad (even easier)".

Only in the small print at the end did Inside Track point out that the leaflet should not be construed to be an invitation or inducement to engage in investment activities, and that the firm was not authorised by the Financial Services Authority to provide investment or financial advice.

Inside Track was not, of course, doing anything illegal, at least partly because there were then and are now few specific rules about property investment by individuals for them to break. Inside Track was not giving investment advice and insisted that the investment decision was left to the client. But for a property journalist, a chance to see the sharpest practice in the business in action was too great a temptation.

I applied to attend a "free seminar" by Inside Track in Devon in the summer of 2006, registering under my own name but not declaring myself to be a journalist.

After my application but a few days ahead of the event, Inside Track repeatedly urged me by e-mail and telephone to arrive 30 minutes early, to allow time for registration. Registration in fact took 15 seconds ("Fill in what you like", said the man with the forms) and for the rest of the 30 minutes Inside Track personnel mingled with the audience, checking what they wanted from the evening and asking about their existing investment portfolios.

At 7.30pm, as about 60 people entered the banquet room, we were told the main doors would be locked throughout the three-hour presentation "for health and safety reasons". The only way out was through a back door, at which were posted five Inside Track personnel. Whether or not this was for a genuine health and safety reason, it meant that leaving the seminar would be an intimidating experience.

We were twice told to switch off mobiles ("putting them on silent or vibrate isn't enough") and a notice said that that no recording equipment could be used.

Then came the evening's entertainment in the shape of a 60-something Inside Track representative. With a deep breath he began an almost non-stop peroration that was in turn populist, bombastic and bullying.

His criticisms were aimed at audience members who folded their arms ("you're just cynics"), estate agents ("middle men only after their cut"), property journalists ("left-wing academics") and broadsheet readers ("if I've upset them, I'm really, really glad").

Those who declined to answer his questions were described as "silly" or "snob-heads" or "losers who haven't got properties of their own". His three-hour monologue — interrupted only by Inside Track colleagues explaining their transformations from financial failures to property millionaires — can be boiled down to four main points.

First, he said attendees should sign up for a £3,995 weekend seminar "where you'll meet developers", "be taught how to structure a deal" and "learn 30 surveyor-based techniques to buy at below market value".

Second, after that weekend, they should spend a further £5,995 to get access to online information about new developments "to cherry-pick the best deals". If they were keen, they could spend £14,000 or more on seminars and meet-the-expert sessions — not a penny of which would actually be spent on bricks and mortar.

Third, if they invested, they should borrow money to buy at least five and preferably 10 or more buy-to-lets, at discount prices secured by Inside Track: "If market conditions change, you've no protection with only one." This emphasised a key message that Inside Track spokespeople were told to push time and again — that investors were seriously exposed with just one or two properties, but would actually be "safer" in a downturn if they had many. This tactic was taken up by other supposedly responsible investment clubs.

Fourth, all deals should be done through Inside Track's linked companies, Instant Access (for property acquisition), Fuel (for mortgages) and Aftercare Solutions (for property maintenance).

To support all this at the seminar there were the speaker's highly opinionated contentions, rattled off with apparent authority: most Spanish properties had quadrupled in value over five years; one Florida villa that allegedly appreciated from $100,000 to $300,000 in 18 months and this was "typical of what's happening there"; and "whatever happens, when you go to sleep tonight you always know your house will be worth more tomorrow."

He finished by giving away a trio of Inside Track DVD sets and challenging his audience to either sign up for the weekend seminars or instead go home and explain to their families why they were "still failures and still broke".

A few audience members stood up, pens and credit cards at the ready, but most of us ran the gauntlet of Inside Track's "consultants" to exit through the only available door. My personal Inside Track experience was over, thank goodness.

A few months before his firm went bust in mid-2008 Tony McKay, Managing Director of the Inside Track group, said his companies' robust style of marketing at workshops should not be compared with how estate agents or developers sell their products. "At these events we're not selling properties but selling the right to attend an educational seminar", he said; it was difficult to tell whether his tongue was firmly in cheek.

Before Inside Track is dismissed as a one-off, remember that around 20 other firms — with less bravado, perhaps — were doing much the same between the late 1990s and 2008, attempting to sell domestic or overseas properties to British buyers.

Inside Track was founded by Jim Moore — who began his business career in pyramid sales schemes — and the firm espoused his philosophy in four crucial ways that typified the investment "get rich quick" bravado of a substantial minority of those at the heart of the UK property boom. It was a characteristic that led, in 2008, many members of the public to welcome the downturn as a means of ridding the property industry of such practitioners and such tactics.

- Inside Track supported extensive off-plan investment, buying apartments for a small down-payment as long as possible before completion. The buyers were then encouraged to flip, making a quick if not necessarily large profit, and then re-invest in still more apartments ... through Inside Track, of course. Some would call this speculation, rather than investment.
- Inside Track urged people to gear heavily, borrowing the maximum amount and leaving them exposed if — and we now know, when — interest rates were increased and lenders' liquidity tightened.
- Inside Track discouraged the spreading of risk outside of property and outside of its own sales and management portfolios. It poured scorn on the roles of independent financial advisors, in part because they advised would-be investors to put money into assets other than bricks and mortar. The result was that thousands put money only into property and are now paying the price.
- Inside Track boasted of, and exploited, the relatively weak regulation of property clubs. Calls from many organisations for IT-type clubs to be regulated through the Financial Services Authority were dismissed by the Government, and of course by the clubs themselves.

In, say, 2006 it was easy to find Inside Track peddling these views but the snake oil sales technique did not end just with that firm. It was not very much harder to find seemingly reputable estate agents in Leeds and other northern cities saying the same sorts of thing about the ubiquitous two-bedroom, two-bathroom apartments that were appearing in these locations by the tens of thousands. Back then, that looked like a business model — many now would think it looked like a cruel deception.

The ultimate insult to those who lost their money because they bought into the Inside Track "dream" was that in early 2009 Jim Moore (who divides his time between Marbella and Geneva, incidentally) was exposed as setting up a vulture fund to buy back the depreciating assets of Inside Track victims. He was looking for investors in a £500m fund to be run by Park Lane-based Sterling Property Assets.

One newspaper quoted Moore as saying, at a dinner at London's Lanesborough Hotel, that:

> Millions were made when the property market was rising, and millions more could be made now it is falling.

Anyone for snake oil?

Conclusion

"Estate agents clearly didn't prepare" insists Henry Pryor, creator of property websites like *Primemove* and the *Register of Estate Agents*. "Most agents were in denial 12 months after the market turned and even after Northern Rock. Few put anything away for a rainy day. Share prices plunged, Humberts effectively went bust. Many found they couldn't survive for as little as six months".

His view has been echoed elsewhere, most influentially in a thorough document by consultancy Key Note, called Estate Agents' Market Report 2008. Its summary put it like this:

> We estimate that estate agency revenues from residential sales amounted to 6.71bn in 2007. Revenues are derived from financial services (such as mortgage and insurance sales), lettings and property management, auctions, valuations and surveys, as well as commission from the sale of property. However, the majority of estate agents' revenue — almost three-quarters — is generated from sales of residential property. The total number of residential property transactions, that is excluding land and commercial property transactions, was estimated at 1.8 million in 2007.
>
> The housing market has been on a roll for 13 years, so a sizeable proportion of estate agency staff have not worked in conditions as difficult as those they are now facing. As a result, they lack the experience required to survive in such a market, and some of the younger estate agents might well withdraw from the market. Indeed, there are already suggestions that the market as we know it today will not exist in 10 years' time, given the currently dire state of the property market and the opportunities for developments on the Internet.

> Given the growing power of the internet, many industry observers believe that, when the estate agency industry does recover, it will not resemble the industry of today. It is expected that a period of restructuring will result in a smaller market and more people will be selling property without using an estate agent.

Do we want the industry to wilt like that, giving way almost entirely to the internet?

Presumably not, for few within the property industry — including those who are independent and not actually estate agents — would doubt that an expert agent serves the public better than sole reliance on technology.

But is it time to embrace more technology and, crucially, change the business model that has driven traditional UK estate agency for well over a century? In the next chapter, we will see how that could be done.

3 Learning the Lessons — Diversity and Modernity

One thing is for sure about the future.

Whether you are a futurologist who believes houses will be transacted online or a traditionalist who believes high street shops are still the primary vehicle for agents' wares; whether you believe prices will bounce back within months or take years to recover their pre-downturn levels, nothing in our industry will be the same again.

Although the dust of the recession is still moving wildly in the air, let alone settled, it seems unlikely that the business model of estate agents can continue as before.

It has always been the case, since the first days of estate agency in the UK, that the industry has stuck rigidly to one formula for its sales business — that is, commission fees x transaction numbers = turnover. But 2008 was the year that formula stopped adding up.

Let us begin with a brief history lesson. British estate agency has changed little since its beginnings in the 17th and early 18th centuries when, according to Alan Bailey's *How to be an Estate Agent*, published in 1992, undertakers who were aware of those who had died sometimes undertook to arrange the disposal of the property of the deceased as well as handling the dispatch of the body.

The process of the sale in those days was imprecise — remember we are in a period before the Land Registry or clear title, let alone internet portals and laser measuring sticks — and often the price of a property was settled not by experts but by an auction for neighbours held a week after the owner passed away. The sale of property did not become anything like an industry in its own right until surveying and estate agency became professions in the second half of the 19th century.

The Royal Institution of Chartered Surveyors began life (as the Institution of Surveyors) in 1868, and was occupied with surveying, valuing and managing real estate rather than selling it. At much the same time, officially designated selling agents set up their own offices, usually alongside lawyers' offices in city centres. Some agents called on professional valuers to assist in pricing homes, while some judged values for themselves, rather like today.

Indeed, it is interesting — some may say depressing — to note how, with the emergence of a true estate agency industry in the late 19th century, many of the principles and practices established then, remain in place today.

The volume of business for the fledgling estate agency industry grew exponentially in the early 20th century. Between 1900 and 1998 the housing stock of Great Britain grew from seven million to 22 million, according to the Office of National Statistics, and in 1968, the busiest ever year for new homes, no fewer than 414,000 were built; of those about 60% were private and mostly for sale, while 40% were public rented (that is, the old "council housing").

It was during this time that the classic estate agency business model was created. One or more high street shop-front offices would be bought, staffed with a small number of people, usually on a modest salary but dependent for much of their income on commission. Viewings and "saturday staff" were handled by part-timers who did not necessarily work in those same offices during the week. Businesses would normally close on Saturday afternoons and Sundays. Property details were printed and distributed to prospective buyers, advertising would be in the local newspapers and sometimes specialist magazines. The financial well-being of this business unit, therefore, existed on sales commissions.

Any additional activity — rental properties, property management, financial services or ancilliary fee-earning work — would be piecemeal and often non-existent. Where agencies did carry out lettings, they routinely would be in separate offices to the sales side.

As the most recent cycle of real estate boom continued through the Millennium, so agents' offices mushroomed on high streets and at the peak of the market in 2006 the UK residential agency business had an annual turnover of just over £12bn, clustered around four types of agent:

- the top end agents selling only "prestige properties", marketing them in the UK but also being internationally-recognised as ports of call for overseas buyers wanting a slice of UK real estate
- mass market national agents, selling just about exclusively to Britons but being recognised across most of the UK — most of their homes tend to be at the mid- to lower-end of the market
- regional estate agents, who were recognised across a swathe of the country, often having large numbers of offices within that region and usually dealing at the mid- to top-end of the market
- the very local agents, sometimes only one branch businesses, many set up at the height of the market and predominantly run by people whose agency experience is short-term and limited. Their property base is low-end.

This profile of the industry, first set out by business consultancy Key Note, is now likely to undergo major change. Many of the very local agents have gone out of business and almost every remaining agency in the country, up to and including the prestige property-sellers, have closed branches and/or made staff redundant.

So will the business model change too? Certainly the context of, and justification for, the business model are changing out of all recognition.

The formula (commissions × transactions = turnover) works only when house prices are rising (thus creating higher fees in absolute terms, without any additional effort by the agent) and when transactions remain buoyant. When they do not perform like that, disaster befalls the model.

However, for over a decade until 2008 the business model not only avoided disaster but appeared to be a godsend. So although commission fees of, for example, Countrywide estate agency edged downwards from an average of about 1.9% in 1997 to 1.66% in 2006 (according to research by online property business *Brightsale*) the absolute income of Countrywide's agents soared.

Why? Well, property prices rose by some 232% on average during that 1997–2006 period, while transactions remained high year on year. As a result, *Brightsale* claims, commissions earned by all UK estate agents totalled around £1.9bn in 1997 and reached £5bn by 2007. In the same 10-year period, the internet revolution took hold and sharply reduced the overall operating costs for agents — after all, they no longer had to mail details to clients, nor even produce full brochures as these were available in PDF format on the web.

But that model received a kicking in 2007 and especially 2008. But further analysis shows that this was not merely a kicking delivered by the downturn — it may well be that, because of wider demographic reasons, the model is now outdated in the longer term, whatever happens to the market.

Clearly, in years to come not only will prices remain muted but transactions appear highly unlikely to reach the high of around 1.2 million property purchases seen in 2005. In addition, owner-occupation is no longer rising; a survey by the Halifax/Bank of Scotland, based on government data, shows that owner-occupiers in England fell by a record 83,001 to 14.54 million in 2007. This was the second successive annual decline and took the home ownership rate from 70.3% in 2006 to 69.8% — the lowest since 1998. The same trend is clearly happening in Scotland and Wales.

This paints a bleak picture for agents' income in future years. Surely there must be a way of operating, encapsulating the lessons from this recession and these demographic changes? Surely there can be a more sophisticated result than simply having fewer estate agents but exactly the same business model? If we are to garner any silver lining from the downturn years, it must be to have a leaner, fitter, more efficient and more popular agency industry.

Two main things — diversification and modernisation — will be the name of the game.

Those that have practised diversification and modernisation in the past (a minority of agents, but including most of the biggest players in the industry) are likely to be healthier in the immediate aftermath of the downturn than those who have not.

But whatever has been done in the past, everyone will have to do more diversification and more modernisation in the future.

Nick Goble of Winkworth says:

> A lot of agents didn't learn the lessons of the 80s and didn't diversify their businesses. Too many agents start in residential sales and feel uncomfortable branching into different disciplines, especially when the sales market is on the up and there seems no need to. Sales bring bigger commissions. What agents don't consider is a falling market and cash flow. Lettings and property management especially are not glamorous. They do however have amazing cash flows and resilience in a falling sales market. Needless to say when the residential sales companies start to struggle they launch head long into lettings in a desperate effort to maintain their business.

He says other, even more short-sighted industry players are those who got involved with property to sell financial products rather than bricks and mortar.

"Halifax was a classic example of this phenomenon" claims Goble. "They were only interested in mortgages and associated insurance products. Apparently a decision was made some years ago not to continue with their lettings and property management departments. This decision was clearly made by someone outside of property."

If you need evidence of how diversification helps, then look at Knight Frank.

Although its image suffered from its reluctance to acknowledge the slowdown in its public statements, the firm's 2007/08 financial year — so up to May 2008 and therefore before the slowdown grabbed the industry — saw group turnover up 17% to £333.9m and profits only slightly down, from £63.6m in the previous financial year, to £59.2m.

The bonus pool was slightly dented, down from £51.4m in 2007 to £46.4m, suggesting of course that Knight Frank's senior staff did not suffer personally in the slowdown as the partners were presented with an average £780,000 bonus each.

Diversity was the key to the firm's success. The agricultural boom benefited the firm in the counties, while in London its residential development team claimed to advise on half of all the capital's key new developments. Its international activities also blossomed as it continued to restrict most of its activities only to the very top-end of the market.

"We have seen another year of sustained organic growth and remain a robust business. We have suitably positioned ourselves to endure the ever-changing markets which continue to impact across the globe. Our balance sheet is strong with net assets up from £69m to £76m. We have produced a good cash-flow performance and increased our core capital base to £10m to recognise the firm's significant expansion", claimed Nick Thomlinson, Knight Frank's Chairman.

Now of course small agencies cannot suddenly diversify into exotic fields such as international holiday homes or consultancy with international developers. But there are practical measures that even single-office businesses can take.

Diversity — first, the rentals market

It is perhaps extraordinary that so few agents considered having a lettings department before the 2008 recession.

In 1988 there were 1.848 million private rented properties in the UK, constituting some 9.6% of the country's total housing stock; by the time Tony Blair entered Downing Street in 1997 that figure had risen to 2.196 million, equivalent to 10.6%. By the end of 2006 the totals were 2.611 million and 11.9%.

Owner occupation also rose in that overall period from 12.661 million in 1988 (some 65.7% of the total number of households) to 15.442 million in 2006, some 70.2%. The big fall was in social rented — council housing, housing association and the like.

The Rugg Review, which looked into the size, influence and future of the private rented sector (PRS) in the British housing market, said there were four reasons behind the PRS's growth in the 20 years before the 2008 sales market recession:

- the Housing Act 1988, which introduced shorthold tenancies and lifted rent controls on new tenancies to make landlordism more feasible
- the glut of properties that became available for rental in the early 1990s following a slump in housing prices in the sales market — this encouraged many home owners who could not sell to enter the lettings market
- the expansion of demand for private rental accommodation by increasingly mobile workforce and student numbers
- from 1996, the availability of buy-to-let mortgages, which were becoming available at a time when house prices were rising and confidence in the stock market as an investment option was low.

Yet despite all this growth in private rental activity, at least proportionately comparable to that of the owner-occupation market, only 10% of estate agency companies in 2007 had a lettings department. Those which did, quite often openly admitted it was staffed by "less capable" staff than those in sales departments.

Quite a bit of the problem has been undiluted sexism, too.

"I entered the industry in the early 1990s in a major London agency and wanted to be in sales but, frankly, that was male dominated. Originally I was told I didn't have sufficient experience but then I researched the background and CVs of those in sales, and found out I was both better qualified and more experienced. In the end it was obvious — I didn't get into sales because I was a woman", says one highly successful estate agent who now runs her own (surviving and profitable) business.

Peter Rollings of Marsh & Parsons says:

> I learnt a lesson at Foxtons when in 1989 we started lettings. From the very first moment I arrived at Marsh & Parsons, I focused on the not quite so profitable but very valuable cash flow business that is lettings. I didn't put second rate people into lettings because they ëweren't good enough for Sales'. I didn't reduce the advertising for lettings and I focused on an incentive package that meant that lettings negotiators and managers were well paid, not the poor relations of sales. Obviously we lived through an exciting and lucrative sales market; however, one must always have a second string of income and this was ours. As I have always said, lettings is a hedge against a sales downturn.

Ed Mead, a Partner in the Douglas & Gordon agency, agrees:

> Rather like bankers, some estate agent thought they'd move into areas that would make them money and disposed of the old fashioned 'bowler hat' disciplines ... which had always been very unsexy and hardly profitable but now produces a good steady income stream. A big mistake; those old fashioned rounded agents will be the one to survive.

There should be plenty of business for those engaged in lettings and sales once the industry stabilises, post-recession. Certainly, the Rugg Review envisages a much larger sector in years to come — and since it reported, the credit crunch has served to bolster likely rental demand from frustrated would-be first-time buyers and disillusioned former owner-occupiers, too.

The Rugg Review also suggests for the future that all landlords should have a licence to let, which could be revoked if statutory requirements are broken. It also advocates that "social lettings agencies" should do more to house people in the PRS. Tax reform is also advocated to encourage landlords to let as a business rather than an investment, with changes to capital gains tax and stamp duty both mooted.

The PRS will undoubtedly expand if these recommendations are acted upon and estate agents would surely be foolish to avoid wanting a slice of this reliable, if unspectacular sector in the years to come.

There are signs that even some of the most traditional elements of the property industry, the trades bodies, are realising that a new business model for estate agents must include lettings. This is why the Association of Residential Letting Agents and the National Association of Estate Agents have recently merged, but the small number of sales agencies bothering with rentals until recently was perhaps the reason that the merger took a long time to achieve — and still leaves many of the older, conservative personnel on the sales side of the business unconvinced.

More diversity — the property management sector

This is the least glamorous of the ancillary work that traditional estate agents can do — the heavy paperwork activities that, for those few who do it, provided enough income in late 2007 and 2008 for their companies to survive while several "sales only" firms went to the wall.

So little is known about this amongst some estate agents so, at this point, a basic introduction is necessary.

Property management arms of those agents dealing with new-build properties would typically have a range of activities including:

- pre-build involvement with developers in the construction and writing of leases (including terms, ground rent, types and frequency of rent reviews, use of the new commonhold lease-type and so on)
- agreement on and collection of service charges from owners/ tenants in blocks
- the selection of a day-to-day post-completion managing agent — very possibly the property management firm itself, or an experienced local managing agent which the management firm helps to select
- liaison with housing associations or other bodies over the implementation of the planning gain through so-called section 106 agreements
- project and build management — that is, project managing the construction work

- liaison with local communities, neighbourhood bodies and conservation groups, as well as formal local authority bodies — this can be a thankless, bureaucratic and time-consuming task, but much of it is legally obligatory and developers often look for help to outsource this activity
- adherence to health and safety issues during construction and running a block or estate of homes
- staff recruitment and replacement
- budget monitoring
- and ultimately handover to the managing agent once a scheme is complete.

Those management firms that go further and actually become managing agents for an apartment block — either from new, or by tendering to take over an existing block — will typically have these responsibilities:

- to monitor and control expenditure for the block
- maintaining standards of repair, cleanliness and decoration in the common parts of the building such as lobbies, windows, doors, lifts, car parks and gardens
- procure building and other insurance
- place and monitor maintenance contracts for communal equipment such as lifts, boilers and access systems
- provide and replace staff for concierge, cleaning and allied services
- manage pools, gyms, spas or other facilities which are increasingly typical in new schemes
- provide a company secretarial service, produce audited accounts and hold an annual general meeting for all residents.

Unsexy? Dead right. But fees for this sort of thing help pay the bills when the superficially more glamorous private sales sector is as dead as the proverbial dodo, as was the case in late 2008 and early 2009 in much of the UK.

After diversifying, try modernising

There is no definitive model for how estate agency could, let alone should, change to embrace more modern working, but online activity is the chief and most efficient way of injecting modernity into a business.

Certainly, the recession spurred a lot of people to promote online alternatives to the high street shop-front way of selling homes that, even in this internet age, still dominates the sales market.

Some figures may prove persuasive here. Humberts — the troubled estate agency now merged with Chestertons but seen by many to have expanded too rapidly in 2006 and 2007, as set out in Chapter two — followed a traditional estate agency business model, pre-merger. It had little lettings, property management or consultancy income and boasted a highly typical "old-style" breakdown of running costs.

Its interim results ending on 31 March 2007, for example, showed that staff costs accounted for 56.6% of its revenue while premises and administrative costs accounted for another 37%. Only 6.4% was left for profit and other costs.

Is this business model going to survive for long with house prices falling and transaction numbers at a low level?

The answer is a big fat NO.

Few have come forward to say exactly how the model will change but there is no shortage of pioneering firms trying to find a way forward. Most sober-minded analysts of the industry believe some estate agency premises will remain but equally, there may be some activities now conducted in a labour- and premises-intensive fashion that will solely appear online in the future.

The various case studies listed in the following sections do not necessarily show the definitive way forward but what they do show is that many people in the residential industry are now at least looking for a new business model.

Modernity — case study: Oliver Finn

This is a single-office estate agency in Chiswick, West London. Its contract for vendors says:

> Oliver Finn's fees are calculated as a fixed fee depending on the service selected. Fixed fees are irrespective of the price finally achieved for the property. For sole agency instructions the fixed fee will be £3000 + VAT for any Flat (Leasehold property) or £6000 + VAT for any house

> (Freehold). For multiple agency instructions the fixed fee will be £6000 + VAT for any Flat (Leasehold property) or £9000 + VAT for any house (Freehold). These fees are subject to Oliver Finn displaying a For Sale/Sold board at the property and being the only agent to do so in connection with this instruction. In any circumstances where we are not permitted to display a For Sale/Sold board at the property or be the only agent to display one, we will make a charge of £1000 + VAT, which will be added to the Oliver Finn fee.

The absolute figures may sound high for vendors outside of London but in an area where the typical home, even in a slowdown, costs some £500,000 the possibility of paying only £3,000 or £6,000 in fees is a bargain. But the key elements to notice are that it is a fixed fee irrespective of property value, and there is a "fine" if the seller does not want a "For Sale" board on display.

"I charge a much lower fee for the same advertising exposure. We do all the viewings and have a board outside. The difference is that the clients are seeing a larger net surplus at the end of the transaction. We charge the fee at the end on a no sale, no fee basis in the same way [as a mainstream agent]. We've slightly adapted the traditional method by adding further financial discount by having a board outside your home. Boards are the main source of enquiry, therefore our job is easier", explains Christian Harper, who runs Oliver Finn.

"This business model is gaining traction as the public are becoming increasingly aware that our industry is not rocket science. The old days of registering in estate agents' shops on Saturday mornings has all but gone. Estate agents offices have now become meeting places and re-enforcement of brands, more than necessary parts of a sales business", he says.

Harper says we are moving closer to the American and Australian systems of going to one place to see the complete market — both of those countries use a multi-listings service, run by agents.

"In my area we have 47,500 chimney pots and not one shop to buy a washing machine. We used to have three — a local electrical shop, a Currys and a Comet. They've all closed due to the public buying cheaper online and going to mega out of town shops. I feel that estate agency is going the same way", he says.

Modernity — case study: WOW

WOW was launched in 2007 with its business model based around what it predicted would be a market downturn.

Rather than be an online agency with no human interface at all, it claims to have 200 representatives around the UK (you can find their identities and locations on an interactive map on *WOW*'s websites, *www.wowproperty.co.uk*) who will handle the usual form of agent's work, such as preparing brochures and handling viewings if you wish. The core fee is a £999 flat rate, payable only if the property is sold.

"The recession plays into our hands as WOW offers a fixed fee, no frills, value for money service. Other estate agents are being seen recently to be piloting the flat fee of £999 but their main differentiator is to 'pay up front with no refund'. *WOW*'s business principles are based on the standard model of 'no sale, no fee' which is proving to be a winning solution", according to *WOW*'s Chief Executive Gareth Robinson.

"We are continuing to sell properties exceeding the current average of less than one per week and keeping our costs low means we can pass the savings to our customers. Our aim is to take in excess of 1% of the marketplace by 2010 so watch this space", says the highly confident Robinson.

WOW operates from the principle that the estate agency industry is comparable to the supermarket sector. In the latter, by late 2008, over 40% of British shoppers had switched from the "Sainsbury and up" status supermarkets to the "Morrison and down" sector, according to a report by market research group Mintel. *WOW*'s business model is based on the same applying to the sale of homes.

It is easy to pick holes in the business model — most problematic of all is the fact that, with a credit squeeze, there are far fewer transactions of any kind, irrespective of the status of agent used by the vendor. Likewise, the low number of agents operating in the market — in Devon, the second largest English county, there is just one at the time of writing — means the firm could not cope with any serious level of demand.

But at the risk of sounding patronising, *WOW* is at least trying something new at a time when traditional firms are not able to present their business model as a lean, successful example of how to sell homes in the future.

Modernity — case study: the Movewithus affinity group

Movewithus is now a decade old but has recently overtaken the likes of The London Office, TEAM, The Guild of Professional Estate Agents and Home Sale Network to be the largest affinity group in the UK estate agency industry, providing services to a loose network of a little over 1,000 agents' offices.

If you look behind the prosaic claims of the firm — "We aim to get closer and closer to the way a modern day individual would like to deal with their property requirements" — there is a genuine attempt to modernise and diversify the buying, selling and moving process here.

As a website, *www.movewithus.co.uk* offers the usual online-style selection of properties for sale. For a vendor, it offers a different kind of sales service: it will select the agent to sell through, and promises to conduct a twice-weekly review of progress with the chosen agent, and provide a weekly written progress report to the vendor. It pledges to vet offers the bona fides of those making an offer and, once a bid has been accepted by the seller, it will sign up a solicitor and manage the moving date.

It says it handles 5,000 sales a year (or at least did before the transaction slump of 2008) and it also offers a range of property management services including probate, estate sale, equity release and assured maintenance "designed to assist clients during sensitive and stressful times" — for a fee of course.

Its founding director, Robin King, has been quoted as saying most traditional estate agents' business models are flawed — and were so before the downturn. The diversity of services sheltered *Movewithus* from the worst excesses of the recession, and allowed it to continue at least partly because many agents themselves failed to adopt similar diverse business models.

"As the cost of marketing a property increases, agents will have to change their model for charging fees. The concept of ëno sale, no fee' will come under pressure and sellers will have to pay a fair proportion of the risk of putting their property on the market", says King.

Critical to the *Movewithus* philosophy is the end of the old physical high street branch. "More than 70% of estate agents' business comes through the phone or the internet and these customer services will become increasingly important" says King. "Just like banks, the property services sector will have to cut costs from its branches and centralise the business by using more technology."

The *Movewithus* modus operandi is to provide a centrally-sourced service to or through an agent, at a cost that benefits from economies of scale, and then allow the agent to add their own margin. This is how it words its pitch for estate agents to use it as a source for Energy Performance Certificates (EPCs):

> We're making it easy for you to take advantage of this month's EPC changes by passing *movewithus* instructions from your landlords and vendors, creating significant incomes for your office through margins.
>
> As you're aware, all landlords are now required to provide EPCs to tenants prior to entering into a letting agreement. Also, any properties previously exempt from the HIP legislation will now require an EPC.
>
> We can provide these EPCs on behalf of your agency, enabling you to add your own margins and create revenue from each instruction. Don't miss out on this easy opportunity to bring in this additional income, while providing an essential service to your clients.

Once the recession ends it will be clear whether those agents who bought into this diversity of services fared better than their more narrowly-focussed rivals. But what stops more agents providing this breadth of services for themselves? And if they did, would they be better placed to survive the *next* recession?

Modernity — case study: other tweaks to the traditional business model

The 2008/09 recession is continuing to make estate agents think about how they conduct their business. Even those agents sticking to the traditional business model believe there are new tricks to be taught to the old dog.

For example, Lauristons estate agency in London tried charging a £999 flat fee for selling a home instead of the 2% to 3% charged by most of its high-street rivals, saving prospective sellers of average priced properties as much as £8,000.

"The market's changing and the current economic climate means it probably won't be the same in the future", warns the firm's Sales and Marketing Director Steve Truman, who says estate agents should join the majority of other professionals in charging an up-front fee instead of working on a commission basis.

In Southport, two estate agents are trying a different idea — swapping houses. Lynn Thompson and Chris Tinsley run separate conventional agencies but have joined forces to try to "match" sellers, who are encouraged to swap properties of equal value, or trade up or down and pay the difference.

Modernity: going online

"Online" covers a multitude of activities but should be clear enough in its main principle. That is, for estate agents, it involves putting into place a business model where the main platform for buying and selling is online.

There are several different online "presences" that an agent can have.

The most obvious (and one which traditional high street agents have had for many years now) is the simple advertising of properties for sale on the so-called *property portals* such as *Primelocation, Propertyfinder, FindaProperty* and *Rightmove.*

Then there are the anti-estate agent *private sales websites* which allow individuals to completely by-pass agents and to instead create their own listings; some private sales websites let you list a home for free, others charge up to £250. These sites have been almost always highly unsuccessful for a variety of reasons — they have rarely attracted a critical mass of properties to appeal authoritative for buyers, they have no real hands-on assistance for sellers who are left to print details and handle viewings.

But the sort of online presence the industry should surely be aspiring to in the post-recession era is that of the *online estate agent.*

This combines giving buyers and sellers the convenience and modernity of an online sales platform but with the high service potential of a traditional agent.

This approach does not have to be the province of the young.

Modernity — case study: James Whitehead

James Whitehead, in his 50s, was perhaps one of the few estate agents to see the recession coming. In November 2007 Whitehead, an agent for three decades and whose patch covers the Blackburn and Ribble Valley areas of Lancashire, gave up the shop-front premises he had in central Blackburn and instead started operating from home.

He did it with extreme professionalism; the redundant playroom extension became a modern office, he had a website built to deliver virtual tours, floorplans, Google Earth maps and fully downloadable brochures. E-mail is his main form of communication, his postal address is a PO Box, he holds meetings at properties and not in an office, his telephone is always redirected to his mobile, and he operates as a one-person business chiefly from a liveried company car.

This is what he told *Estate Agency News* about one example of his approach to selling homes as a modern business:

> I outsource the production of photography, virtual tours, floorplans and so on, to a company called Rooms PR who charge a nominal setting-up fee with a balance to be paid on successful completion of a sale. It's a brilliant service which takes away all the production hassles. Everything gets uploaded to my website and is compatible with all major portals. Again, this saves staff administrative time. Software tells me that my website is very active, my instructions are constant and even in the present climate my sales are positive. However, the saving on overheads has been phenomenal. If you are an experienced estate agent with an established name and a reputation for getting the job done, there is nothing to stop you doing the same.

Whitehead saves between £38,000 and £40,000 on overheads — in addition to no office and no admin support, he has ended all his local newspaper advertising — and believes he offers a better service to boot.

But one of the most fascinating aspects of online activity is that it does not require just one template that every agent must follow, as these case studies show.

Modernity — case study: SimplyZigZag

Launched in June 2008, this service extends the traditional cheap and cheerful online agency principle into something that offers at least a few of the advantages of a "physical" estate agency, too.

For £50, you get a three-month listing with five images, plus password protected access. For £500, *Simplyzigzag* acts as a full agent, handling enquiries, organising viewings, and — most importantly — earning access to a number of portals that cater to agents-only, including *Rightmove* and *PropertyFinder*. For £1,500, your property is featured on the website's home page, you get a dedicated account manager who'll oversee the entire process, from offer to closure, to liaising with the solicitor.

A weakness for the seller here is that, unlike traditional estate agents and many other online offerings, this is not a "no sale, no fee" service — a winning appeal in recessionary times.

Modernity — case study: Brightsale

In the North West of England, Thornley Groves, a "traditional" estate agency chain, has bought 25% of *Brightsale*, a pioneering online estate agency. *Brightsale* is not a free-to-use website that leaves viewings and sales negotiations to sellers, but calls itself a "proper" estate agency — it simply has no physical branch. Its trained negotiators instead advise sellers on how to handle buyers, and will arrange floor plans, photographs and brochures. The fee, if a home sells, is 0.5%, or only a sixth of the levy by many conventional agents.

Many property analysts think high street agents will have to adopt a similar business model after a recession in which many sellers may have "done it themselves" to save fees — in which case, Thornley Groves is at the start of something big. "We'd been searching for a way to bring estate agency into the 21st century. Clients love what we do but in the internet era they expect more. Estate agency has a long way to go to harness the power of the internet for the benefit of its customers", says Thornley Groves' Managing Director Michael Groves.

Modernity — case study: Globrix

This is a curious one, a property portal that simply uses software to scour estate agents' websites for property details, and them assembles them in one place — that is, *www.globrix.com*. It then allows the public to scour for properties by keywords.

"You should also know that we don't accept paid placement or listing fees from estate agents. They can't buy their way into our search results. We're completely independent and unbiased that way. Instead, we make our money by only allowing targeted and relevant ads down the side of the page", *Globrix* says.

Well, it is slightly more complicated than that. Backed financially by Rupert Murdoch's News International, the site has had free publicity from Murdoch's papers — one embarrassingly fawning article in *The Sunday Times'* property supplement drew well-deserved derision from *Private Eye*.

But, clunky publicity apart, what is strategically important about this website is that it shows that multi-national media interests outside of the property industry have seen the potential of investing in an online property portal.

Agents do not have to pay to promote properties, so have welcomed it with open arms (and have used its free-to-advertise principle to threaten the very existence of rival portals which charge agents, like *Rightmove*, for example).

Modernity — case study: Lettingweb and Facebook

This is not operating a real online estate agency model but still has an interesting innovative approach from which existing agents can learn.

Lettingweb has been established for some years, advertising individual and institutional landlords' properties to rent and concentrating on big cities in England and Scotland.

At the height of the recession, it launched an application on the social networking website *Facebook* aimed at winning more tenants for its advertisers. The application, at *http://apps.facebook.com/lettingweb*, allows users to search for available rental properties on *Lettingweb*, make inquiries directly to the agent and share property details with their friends — a distinctive characteristic of *Facebook*.

Alex Watts, Director of *Lettingweb*, says:

> We are 100% focused on delivering tenant and landlord enquiries to our member agents. The average age of UK renters is 16 to 35, which is the same demographic as the majority of *Facebook* users. So it made sense for us to target the site, making it quick and easy for users to see our property content and discuss available properties with their friends and potential housemates.

This is an example of using increasingly diverse online activities to target audiences.

Modernity — case study: how the US uses multi-listings services

It seems absurd to use the word "modernity" for this, as it has been operating in the US for over 20 years now (and in many other countries for over a decade) but at the core of the American realtors' drive to modernity has been the multi-listings services (MLS).

This system allows open access to property sales information across rival estate agencies. A seller registers with one realtor in an area and is told that the house details will be made available to all other realtors subscribing to the MLS — in the US, this tends to be the majority or often 100% of realtors in a locality. Likewise, a buyer who registers with one realtor in that district is told the names of rival realtors in the MLS to prevent any duplicate registering.

An early example of this in the US actually started back in 1989 but remains the largest such system. It is the Multi-Listing Service of North Illinois (MLSNI) which in 2006 (the final year before the large scale US housing market downturn) had no fewer than 75,000 properties for sale on its database, or the equivalent of some 85% of the homes on the market in that location. It contained details of a further two million properties not currently on sale, plus over 600,000 photographs of properties, as well as the public council tax records of some four million flats and houses.

So with little effort on the part of potential buyers, they can look at comparable properties, sale histories, a rough visual condition of their chosen home and an idea of how much it will cost to run.

At the height of the pre-slowdown US real estate market, the MLS system across much of the US was owned by just 10 agencies or networks of agencies. In total, this represented a startling 35,000 realtors, ranging from big corporate brands to individual sole traders.

New technology would make the introduction of this kind of system in the UK very straightforward. In a post-recession world, where perhaps many sellers may be wary of estate agency fees, something like a MLS might convince them that using an experienced, modern, high-tech agency is better than going it alone on some private sales website.

Conclusion

Shortly before the recession, Steve Rist — who used to run the well-publicised but ultimately unsuccessful private sales website *Easier.com* — went on record as saying that a combination of a backward-looking estate agency industry, a recession and newcomers like supermarket house sales, would threaten the status quo. Rist said:

> What amazes me is that estate agents have apparently done nothing to meet this inevitable challenge. And it is not as if they hadn't already had a warning shot across their bows. It was Valentine's Day 2000 when I floated *Easier.com*, a self-sell website for homeowners loathed by the

> traditional high-street agent but loved by sellers. At its peak, so many people were flocking to list their homes on it that the total value of the properties contained in its database touched £3.5bn. The trouble was that the site was years ahead of its time and, when the dot.com crash came, we ran out of cash. After Easier failed, estate agents put their heads back into the sand. Rather than develop their business model to prepare for the next onslaught, they stuck to the old commission-based structure and, as a result, are in real danger again of pricing themselves out of the market.

Easier.com was perhaps destined to fail as it offered little practical support for a seller, beyond being a cheap vehicle for publicity. Yet Rist makes many good points.

If the estate agency industry does not reform itself, there are many outside ready to do it instead.

Just as Rupert Murdoch's *Globrix* has taught *Rightmove* a lesson, so perhaps Asda or Tesco, notwithstanding their faltering steps so far, may one day go on to teach Reeds Rains, Connells or Your Move a lesson. It sounds preposterous now, doesn't it? But then again, the concept of nationalised banks sounded preposterous just a short while ago.

One way to at least fight back and move forward would be for existing agents at the mid- and lower-end of the market to take a leaf out of the books of their well-to-do rivals, and stabilise their business model by diversifying the property-related work they do, such as lettings and property management and consultancy for local developers who ache to have professionals take management chores from them.

Then the second step would be to consider reducing the unnecessary overheads by relying more — not wholly, just more — on online activities. This is especially appropriate for those dealing with large volumes of similar properties, which lend themselves to greater levels of internet-driven marketing.

Let the top-end agents with their fat commissions, their bonuses and their clientele who are partially insulated from the slump, continue with expensive offices and labour-intensive techniques. But for the volume market, take up some of the lessons of those case studies above, and move towards increasing technology and reducing running costs.

Put like that, it sounds just plain sensible, a sort of short-term survival exercise. But it could herald a revolution in the agency business model ... and it might move the whole industry into the future, too.

4 Disaster for the House Builders

2008 and 2009 should have been busy years for British residential developers, who not so long ago would have reasonably expected to be hard at work fulfilling the targets set out by Gordon Brown as recently as July 2007.

Back then — and it was so recently, despite now sounding as if from another age — Brown announced that in England alone there should be between 240,000 and 297,700 homes built annually until 2016, and that, in total, between 2.9 million and 3.5 million new homes should be constructed by 2020. In addition, all new homes were to have a zero carbon rating from 2016 and there were to be 10 new eco-towns by 2020.

The carbon target may yet be met as the Government's Department of Communities and Local Government has continued sending out directives on how to achieve the goal throughout the downturn. But the dramatic falls in completions has perhaps been the most potent symbol of the reversal in fortunes of the housing industry.

In 2007 — before the downturn hit the new-build market — some 174,900 homes were completed: still below target but on an upward path from previous years.

Yet in 2008 scarcely 120,000 were built, according to the House Builders Association. It predicts 2009 and 2010 figures may well fall to dismal five figure levels. The consequence is that the new homes industry is imploding. The fewer homes built, the fewer people work in the industry, making the downturn even worse. Similar targets, and shortfalls, exist in Scotland, Wales and Northern Ireland.

Jim Ward of estate agency Savills predicts new-build levels in England will not have returned to 140,000 per year — scarcely half the Government's target — even by 2013. He says:

> Once sites are mothballed, there's inertia in the system as it takes time to rebuild teams and return to the same master-planning position on more complex sites.

Measures announced to kick-start elements of the housing market throughout 2008 did little to improve building rates. Stamp duty changes may have created a marginal boost to sales of old stock but did nothing, it seems, for new building rates. Allowing housing associations to buy unsold houses and flats may in theory mop up stock but it does not help overall supply meet demand — and in any case, the associations have not bought many private homes, claiming they were dominated by too-small flats with poor build-quality when what they really wanted was larger homes that were built to last.

The slump in building, of course, has not just affected the private sector.

Many of the affordable homes made available to key workers and the low-paid are built as a result of so-called "section 106" deals — that is, arrangements by which developers have to build a number of affordable homes to get an agreement to build private-sector properties. With private sales at a standstill, developers have downed tools on residential property across the UK, so section 106 properties are not being built either, creating a shortage of properties exactly where demand for affordable homes is strongest.

In addition to the slump in actual building, a near-total collapse of demand from individual purchasers, severe restrictions on borrowing from institutional funds, plus the start of a general economic recession, 2008 was also the year that residential developers in the UK had to start writing down their asset-worth because of an unprecedented drop in the value of land.

Urban developable land outside of London and the South East had fallen in value by around 35% by the end of 2008, according to the research department at Knight Frank. Greenfield sites fared only slightly better with falls of 30% — but faster falls occurring at the end of 2008 suggested the worst was still to come.

Yorkshire and Humberside had been hit hardest by the downturn, with land in both brown- and green-field categories falling 50% in 2008. The North West of England was close behind with drops of 41%

and 36% on brown- and green-sites respectively, while prices in central London dropped a more modest 10% and outer London by 15%.

"Developers have put their land acquisition activities on hold, which has dramatically reduced demand for sites — by as much as 60% in some parts of the country. They've found it almost impossible to access finance to buy land, while the slowdown in the sale of new homes has prompted them to reconsider their future needs. Indeed, many are selling sites to raise cash and bolster their balance sheets which has dramatically increased the supply of land on the market, further depressing values", according to Jon Neale, Chief Author of the Knight Frank research.

"There's evidence that many other vendors have not yet come to terms with the changed market conditions and have unrealistic expectations of what price their site can achieve, particularly if it was bought [by them] at the top of the market", he says.

Neale believes only so-called "oven-ready" plots in prime locations with existing planning permission attract any real buyer interest, but much of the available land does not fit this description.

And if proof were needed of the change in the residential building industry, Neale says that outside of London some 30% of all developable land had in 2008 been bought by housing associations — and just 16% by private sector residential developers:

> More supply will come to the market as developers reduce the size of their land banks, and it will be some time before they are ready to acquire again. Values are likely to continue to fall, albeit at a lower rate of around 10%.

Little wonder that private developers wrote down their assets to the tune of some £3bn in 2008 alone, with further falls in 2009.

Barratt

Barratt Homes was at the forefront of the price-slashing developer tendency that appeared during the slowdown — although the firm tried to keep this secret from its rivals and journalists, presumably to keep some confidence amongst wary shareholders.

The company had acquired Wilson Bowden in April 2007 for some £2.2bn in what was the first major acquisition by Barratt, the largest acquisition to date in the UK new-build sector, and the largest ever purchase of a UK house builder. The acquisition was engineered by the then new Barratt Financial Director Mark Pain, who in his previous job at Abbey presided over the writing off of £256m in its business banking unit.

By 2008, Barratt had a network of 35 house building divisions throughout the UK under the Barratt, David Wilson and Ward Homes brands. It had written down the value of its land by £208.4m by late 2008 and had received a large amount of (apparently unwanted) publicity in the UK and overseas by offering discounts of over 40% to people wanting to buy five or more of its units on some schemes, especially those in the Midlands and north of England.

It was hard to keep a lid on the straitened circumstances within the company and figures produced by the firm at the end of 2008 showed how dire the market was, for it and rival developers.

Visitor levels per site for the 19-week period ending early November 2008 were down 5.8% compared to the same period of previous years; net reservations averaged 197 per week, equating to 0.36 private sales per week per site, down 23% on the previous year; cancellation rates averaged 24% for the period, compared to 23% in the previous year. The same period in 2007, of course, was already relatively weak as the new-build residential market was by that time already seeing a drop-off in demand. By Christmas 2008 the firm was busy shedding what had become unwanted assets in the shape of an office block and commercial property inherited from Wilson Bowden.

Barratt's quick-off-the-mark price cuts were hinted at by journalists from as early as the spring of 2008 — although Barratt public relations officers attempted to rebuff the claims at the time — but Alastair Stewart, an analyst at Dresdner Kleinwort, blew the firm's cover more comprehensively in October.

He returned from a tour of Midlands and northern English city centres where he spoke of the over-supply of new apartments leading to housing market "carnage beyond even our most bearish expectations." Part of the wreckage he saw was Barratt's deal of a 43% discount off deals of five or more flats in East Yorkshire.

Suddenly, it was clear how bad the market and Barratt's position were — and the PRs stopped giving the journalists a hard time over it.

It has been a baptism of fire for Mark Clare, who was Head of Residential Business at British Gas and took over from retiring veteran David Pretty as Barratt Chief Executive just as the company recorded 14 consecutive years of growth — until the crunch came along.

Taylor Wimpey

If one developer made Barratt appear a relatively happy ship, it was Taylor Wimpey. This is a firm which managed to have a portfolio of house building simultaneously in the UK, US and Spanish residential markets — three of the world's most recession-prone markets, according to analysis by the *Economist* magazine.

In the UK it traded as George Wimpey, Bryant Homes, G2 and Laing Homes, and by late 2008 its figures made Barratt's appear almost buoyant.

Taylor Wimpey said that in the third quarter of the year prices had fallen by 14% on a year earlier, with net reservations 27% below the 2007 level. "We expect to maintain prices at current levels for the next few weeks" was one particularly desperate sentence in an interim trading statement, indicating the short-term thinking prevalent in the company at the time.

It had 400 sales outlets, down from 500 in January 2008, had an order book of 6,607 homes (down 11,074 homes a year earlier) and had reduced staff headcount by 1,900 in nine months.

The big issue for Taylor Wimpey is its extraordinary level of debt, a fact that many believe may yet take the company under. The outcome of the debt restructuring remains unknown at the time of the publication of this book — suffice it to say that with an eye-wateringly vast £1.9bn of debt, it is seeking to revise its covenant structure.

Bellway

By the start of 2009, the Newcastle-based Bellway group appeared to be producing more statistics about the downturn than homes. To its credit, however, it was transparent at a time when some developers were keeping their activities cloaked in secrecy, from their own staff and investors as much as from anyone else.

In early 2009, Bellway admitted it was selling a maximum of 60 homes in an average week, over 50% below the level it enjoyed a year earlier. Over 95% of its sales were based on incentives, including cash discounts up to a reported 25%. Its order book stood at £340m down from £677m some 12 months earlier, with cancellations running at 24% — an all time high for the company — and visitor traffic down 50% on the previous year. The company wrote down its land bank in late 2008 by £131m or 8% of the total value of its holdings.

In early 2009, just as developers could have benefitted from positive attention, Bellway scored something of an own goal by choosing to pay its three senior managers £632,500 in bonuses. It had been part of what was supposed to be a bonus system based on the firm's performance for the year to July 2008. The firm argued that it had performed strongly compared with Taylor Wimpey and Barratt — indeed it had, but even so, profits fell 30% in the period for which bonuses were delivered.

Persimmon

Land value write-downs were the key problem for Persimmon too — although there were plenty of other challenges. In 2008 there were £640m of land write-downs, with prices down 10% and build rates collapsing by 35%.

"The uncertainty created in the housing market by the increasingly turbulent and uncertain outlook in financial markets has had a negative impact on all our regions across the UK", the company said. "The availability of mortgages has continued to be very restricted and cancellation rates have increased to about 35%."

Bovis Homes

This firm had always been one of the most "straight-laced" of developers with lower than average leverage and a spread of property types rather than concentrating on apartments in recent years. As 2009 dawned, it announced it had a £220m banking facility to safeguard its activities until 2011.

It said the new facility would include a "covenant package more appropriate for current trading" at the same time as increased pricing from the banks to reflect their own new trading conditions. In an interim statement issued six months earlier, the firm admitted that it had stopped buying land to conserve cash.

Meanwhile in London ...

Few markets — perhaps with the exception of the apartment-riddled city centres of northern England — demonstrate the sharp reversal of fortunes of the development market.

Analysis by King Sturge shows that in 2008 there was a 57% fall in new home starts across London, from the 2007 total of 28,800; fewer than 1,000 homes were started in the second half of the year according to King Sturge Research Director Guy Weston. In prime central London — the term given by property insiders to the most expensive areas of the capital — the number of units started did not reach three figures.

This was despite evidence to suggest that some residential development land values in the capital fell by only 15% in 2008 — by no means as much as elsewhere. But even so, that fall meant bad news for the likes of Nick and Christian Candy, the interior designer brothers whose high profile development schemes for the world's wealthiest buyers and investors were making waves from 2004 onwards.

The Independent newspaper saw the brothers' rise to fame like this:

> Nick and Christian Candy bought their first flat in Earls Court in the mid-1990s with £6,000 loaned to them by their grandmother. At that time, one brother was an investment banker, the other was in advertising. They did up the flat, laid a new carpet, sold it, and bought another. In 1999, they founded the property company Candy and Candy, which now has slick headquarters with banks of employees designing sample interiors. As they progressed, they specialised more and more in dream homes for the ultra-rich.

But as the brothers' ambitions burgeoned, so did their gearing and exposure to risk.

Project Blue, a joint venture between the brothers and investment group Qatari Diar, reportedly paid the Ministry of Defence almost £1bn for the Chelsea Barracks site in late 2007. The Candys hoped to sell 320 private flats for £10m each to foreign buyers, but the idea — part of a bigger plan for 638 homes, a sports centre and hotel — hit local opposition as well as sharply falling demand created by the credit crunch.

The brothers have radically reconfigured the Chelsea plan to include a park — although critically its financial underpinning still relies on substantial private sales. The pair have also withdrawn as construction partners in another controversial scheme, Noho Square in Fitzrovia, which involves Kaupthing, the troubled Icelandic bank which has been nationalised by the Reykjavic Government.

Although many of the Candys' rivals looked on smugly as the brothers became increasingly mired in complicated reversals from deals, they knew that in many ways the brothers' problems meant even the top end of the residential market — previously immune from the financial crisis — was now as vulnerable as any other sector.

"The prime central London residential property markets have historically been sensitive to the fortunes of the financial markets as they are highly dependent on demand from those employed in the City. It should come as no surprise that during the latest period of unprecedented volatility in the FTSE and uncertainty in the banking sector, prices and transaction numbers have continued to fall. Rents have also been put under downward pressure for the first time since 2001 due to the withdrawal of corporate tenants", explains Savills' Research Director Lucien Cook.

"Although the Government's measures to bring some liquidity into the financial markets and secure the capital position of the country's major lenders is intended to underpin the future of the banking sector and resuscitate the mortgage market over the medium term, the prospect of significant redundancies and reduced bonuses will continue to curb demand for prime property. We expect that values will fall further", he warns.

Across the capital in King's Cross, apartments in the £2bn redevelopment of the railway station and surrounding area — one of the largest redevelopment projects in Europe — were almost a casualty of crisis.

The eight million square foot residential, retail and office scheme is now being funded directly by Argent, the developer, backed by the BT pension scheme; a consortium of banks behind it, including EuroHypo and Deutsche Postbank, dropped out — at least temporarily — when the money markets went into shock in late 2008. So the scheme goes ahead but if Argent hits problems, it represents not only bad news for the firm and the scheme but also for those with BT pensions now or in the future ... a sign of the inter-relatedness of so many people's fortunes in this recession.

... and everywhere else, too

Needless to say the residential development industry suffered across the country as well as in London.

Gregor Shore, the luxury Scottish developer best known for the Platinum Point and Grantham Marina schemes in Leith Docks, went into administration. Magellen Residential, a builder of schemes in Leeds and Bradford, went under too. The £250m redevelopment of the troubled

Brighton Marina site — a large retail and residential area a mile from the city centre — was put on hold by Brunswick Developments. In the Midlands, Wolverhampton-based Bentley Homes suspended work on its schemes — more flats, of course — and went into administration; the same fate awaited K W Linfoot, Keren Homes and its subsidiary Keren Birmingham. Asquith properties, a West Yorkshire-based developer, went bust as well.

Builders tried various tactics to ease the pain too, the most popular of which was renting out those properties that were completed but unable to attract buyers. Most did it with a pledge to regard the rent paid as a deposit on the property if and when the tenant chose to buy — a vital source of cash-flow for the developer, who then possibly avoided slashing prices when selling the property on, as the "rent deposit" could already be seen as an incentive.

However, while that was a short-term palliative it had a dubious edge to it. Even before the recession created much higher unemployment — which reduced demand for private rental properties — gluts of unsold apartments being put up for rent made local lettings markets suddenly take a hit.

With the lacklustre response to developers' let-to-buy initiatives, the builders must have noticed how few friends they had in the property industry. But a far bigger problem occurred when private developers turned to residential social landlords (RSLs) to bail them out.

The RSLs say no

Not only were the developers in trouble over lack of sales, toxic debts and increasingly pessimistic forecasts for the economy in the UK, they also had the major problem of vast numbers of unsold units being completed even as the credit crunch was still unfolding.

But who on earth would buy them? Richard Donnell of business consultancy Hometrack, says: "There's £10 billion of stock under construction and not yet legally complete." Some market analysts believe that could mean 40,000 new units still awaiting owners.

Savills' research showed that in London alone there were 1,800 empty new units at the end of 2008, with another 1,400 expected by the end of 2009 and a further 1,000 a year later — so 4,200 in total by late 2010. Of course, numbers thereafter will tail off rapidly as many builders are mothballing or simply scrapping schemes that are not already beyond the foundations stage.

But with that scale of unsold homes built or part-built, it is unsurprising then that developers had been pro-actively contacting RSLs — housing associations as they used to be called — tempting them with discounts.

"We've been inundated. Almost daily there are letters or calls from developers talking about reductions for bulk sales", says the chief executive of one RSL who did not want to be quoted directly.

"The real problem is that almost all of the private units completed now or in the next two years, in London and elsewhere in the country, will be two-bedroom, two-bathroom units. The majority of registered social landlords want larger ones, typically three or four bed family units", says Marcus Dixon of Savills' research team.

Dixon says that irrespective of the unit size, RSLs could probably secure a 25% discount off a lender's valuation price of a new-build unit in today's market. As lenders are insisting their panels of valuers produce second hand resale values, excluding the new-build premium, that is 30% to 35% off the "old" price of units. And as we have seen, Barratt and others are happy to go still further.

That sounds a virtual steal for RSLs but they were being warned off by most of their independent advisors. "RSLs pick up hefty service charges that come with the private apartments, so that burden for 50 years ahead could easily wipe off a discount", cautions Savills' Dixon.

As a result, many housing associations have been distinctly lukewarm about the idea.

Geeta Nanda, Chief Executive of the Thames Valley Housing Association, is typical of many RSL heads when she says:

> There's nothing wrong with the idea in principle. It suits some RSLs, but not ours. Running costs are likely to be too high. In any case, we're in this for the long term so we need a more fundamental approach rather than buying into a development which has different parts with different owners.

Nanda, whose RSL handles over 5,000 rented units and 4,000 shared ownership homes in London and the Thames Valley basin, says RSLs should not have their roles moulded by the need of private developers to sell stock and improve cash-flow.

"We should consider taking advantage of opportunities to buy land, develop our own stock, be less committed to old ways of doing things. The days of dense flatted schemes may be ending", she insists.

Some RSLs were keeping their powder dry, waiting to see if new-build prices fall further and discounts rise as a result. Two, however, have decided to bite the bullet.

Sanctuary Housing — a major RSL managing over 70,000 homes in England and Scotland including rented, sheltered, student and key worker accommodation — bought 335 units from Bloor Homes on 19 sites in the first deal taking up some of the Government's accelerated funding. Bloor, Sanctuary and the newly formed Homes and Communities Agency then went into discussions to extend the deal further. Ken Allen, Head of Affordable Housing at Bloor, said in a government press release: "We've been most impressed with the speed of decision making and the approval process."

Colchester-based Colne Housing, which manages 2,000 homes across Essex, saw things differently. It bought 30 properties, mostly houses, from one unnamed developer, but Colne chose to use its existing funds rather than seek funds from the new corporation pot.

"The £200m deal is created for bulk purchases of hundreds of units across a whole region or major urban area. We've instead looked at individual opportunities to buy in line with our overall strategy and demand. We need more flexibility. We've said no to a large number of offers and units, mostly because they were flats in the wrong place", explains Mark Powell-Davies, Colne's Chief Executive. One of Colne's bulk purchases was of a small block of apartments which allowed the association to use its own facilities-management services, rather than buy into high priced private developer service charges.

On eco-issues, most of the private-sector flats looked at by Colne did not reach the association's preferred build-standard, which it enshrines in its own Incini social housing development arm, according to Powell-Davies. However, he says the scale of the discount meant that it was worthwhile buying these units.

Colne's experience was typical of many housing associations. The irony is that as RSLs are generally unwilling to buy much of the flatted product unsold by the private sector, desperate developers are now poaching RSLs' clients.

"Developers are taking on RSLs at their own game by selling some of their private stock on a shared equity basis, simply to improve affordability for purchasers and to maintain cash flow", notes Anthony Lee, Head of Affordable Housing at Atisreal.

"RSLs should be very worried about this. When it comes to choosing between affordable housing product provided by an RSL or a developer, RSLs still have to get over the 'branding' issue that doesn't affect developers", he claims.

But both the public and private sectors say that, whatever has happened so far, all is up for grabs as the recession bites. The Government is increasing funding for RSLs to buy some of the ever-increasing private surplus — what a pity, perhaps, that the private developers' business model did not suggest there was a lack of demand from the core private clients to begin with.

GuestInvest and The Beach — the nadir of pre-recession developers?

If Inside Track was the bastard child of buy-to-let greed, then the idea of building hotels to sell on a room-by-room basis was perhaps the nadir of good sense in the pre-recession development industry.

Therefore, many in the property industry were openly delighted when GuestInvest, run by the flamboyant Johnny Sandelson, went bust in 2008. He did not help himself by making a publicity splash when paying £120m for two central London properties for conversion to hotels in January 2008, at the very top of the market.

The concept — to sell a room for as much as £300,000, give the investor 52 nights a year free accommodation and some form of rental return — was riddled with problems for investors.

First, there was no established resale market to prove if hotel rooms appreciated over time, even when the mainstream residential market appreciated. "Studies in the US suggest capital appreciation will mirror the mainstream housing market", said David Galman of Galliard Homes, a buy-to-let developer who tried his hand at hotel rooms. If he was right then, that means many hotel room investors are in trouble now.

Second, it was very difficult for investors to get mortgages. "There's not much appetite among lenders for this investment, which, in the past, has been considered riskier than conventional buy to let. We have a pool of lenders, among which there used to be three willing to give mortgages on hotel rooms. Now they've withdrawn those deals", said Melanie Bien of Savills Private Finance, a mortgage broker, in early 2008. Unsurprisingly, no lenders have stepped up to take their place as the crunch has continued.

A year before he founded GuestInvest in 2004, Sandelson was also the public face of Ampersand, a developer that was planning to launch The Beach — a scheme in Carlyon Bay, Cornwall — as Britain's first "sale and lease-back" investment resort.

He promised 511 apartments, a hotel and other tourist facilities would be built at the beautiful bay. Investors would buy flats, have free use of them for part of each year, then lease them back to the developer in return for a rent guarantee.

Ampersand ambitiously priced two-bedroom flats at The Beach at £280,000. Buyers would pay 25% of rental income to cover hotel management fees, a £3,500 annual service charge, £300 ground rent, £1,500 for cleaning and £1,000 for contingencies. This would take a large chunk of the £36,650 that Ampersand predicted each investor would get annually, based on letting the apartments for 31 weeks per year.

Critics derided the scheme's viability, attacking the optimistic assumption that people would pay top dollar to stay on an exposed beach in Cornwall outside of summer. The scheme then hit planning problems, and immense local opposition. To this day, the development remains unbuilt but the beach in question remains a sordid building site.

The problems with GuestInvest and The Beach are tiny footnotes in the history of developers' problems in the great slowdown. In any case they were, in some eyes, largely self-inflicted; this is quite unlike the larger-scale traumas suffered by the volume developers.

But the idea of building hotels to sell on a room-by-room basis, and a scheme predicated on 31 weeks of holidays a year on a north Cornwall beach, show how even some developers lost track of what was sensible.

Conclusion

The majority of house builders construct good quality homes having assessed demand, try to price sensitively, and offer good after-sales care. They provide a genuine public service.

But it is not hard to see that a minority of the development industry — not necessarily a small minority — were either blissfully unaware of some of the problems they had a hand in creating, or were fully aware but thought the downturn would either not occur at all or miss them when it happened.

The problems were over-supply, insufficient build quality, wildly over-optimistic business plans and a simple hope that buyers would be gullible and buy anything made from bricks and mortar.

They do not combine to paint an edifying picture. Nor do they suggest the developers' business model is a sound one, given the glut of builders which went to the wall (or nearly did) at all levels of the market during the downturn.

After the recession ends, what more sensible approaches should developers take?

5

A New Business Model for Developers?

Siren voices at the depth of the house building recession in 2008 suggested that whatever happened to the market, we were past the best of the industry.

They claimed that there would be no return to what they were already seeing as the halcyon days of 2000 to 2007. Well, it all depends what you mean by halcyon days ...

In reality, the UK building industry had far more productive and, some would claim, more innovative periods than the start of the 21st century.

Build levels had been far higher in past decades than in the 2000s, architecture had arguably been much better in classic Victorian and art deco periods, although few would doubt that build quality was superior in recent years than in, say, the 1980s — yet build quality might reasonably be described as no better, or perhaps even worse, than in the Victorian and inter-war years which saw the completion of millions of houses which remain popular and in good condition today.

In reality, the British residential construction industry has always been in a state of flux and the latest slowdown may well have shaken it up and re-shaped parts of it. But this is just the latest chapter in a book of events that have moulded the industry.

In the late 19th and early 20th century, house building as we now know it did not exist — no one firm constructed more than about 400 properties in a year (usually far fewer) and these were usually in one region rather than across the whole country.

Modern volume house building began almost a century ago when the 1923 Housing Act provided cash subsidies for the construction of houses for the working classes. The subsidies did not last long — under a decade — but builders responded with enthusiasm, and by the start of the 1930s as many as 200,000 homes a year were being built, although again by regional rather than national companies.

But some regional builders did start moving to London where Costain, Henry Boot, John Laing and Taylor Woodrow became prominent and were a few of many construction firms which created 1,000 or more homes per annum. World War Two and the mixture of austerity, rationing and government controls put volume building on hold until the 1950s — and then the industry let rip for two decades.

The Calcutt Review of 2007 carries a useful appendix covering, amongst other things, this period of building work. Not only does it show the explosion of building at the time but it also reveals how many companies there were in the industry. It says:

> In 1951, output had hit a post-war low of around 23,000 units; by 1954 it had recovered to 93,000, and by the second half of the 1960s consistently exceeded 200,000 units a year.
>
> By the beginning of the 1960s, the top ten housebuilders were producing some 14–16,000 houses a year, not far short of the level in the late 1930s and, at around eight to nine per cent, a higher share of the market. However, with the exception of Wimpey, which stood head and shoulders above its competitors, the larger pre-war house builders had not been overly successful in restoring their earlier volumes; a newer generation of house builders was emerging.
>
> By the end of the post-war boom, the early 1970s, there were some half dozen companies building more than 2,000 houses a year and as many as 26 building at least 1,000 a year; the top ten were responsible for 32–33,000 houses a year, and their market share had risen to around 17–18%. Perhaps of more interest as an indicator of corporate change, of the top ten house builders in the early 1970s, only one (Wimpey) had been of any significance before the War — five of the six largest firms were post-war creations — Northern Developments, Whelmar, Bovis, Barratt and Broseley.
>
> It was in the 1960s and early 1970s that we saw the development of a number of semi-national house builders to go with the one undisputed national. Before the war, there was no doubting that house builders were almost entirely localised concerns, with no more than a handful building sporadically outside their home area. In contrast, by the end of the post-war boom Wimpey was operating nationally through a regional structure; on a smaller scale, Bovis Homes also claimed a national organisation.

More common was the medium-sized firm that covered a significant area of the country — Northern Developments and Barratt, for instance, building across the north of England. By 1973, the regional house builder that we now know had arrived, helped in part by stock exchange financed acquisitions.

The crash of 1974 saw an end to the period of growth. The FT Index fell 55% in a year, secondary banks which funded much residential development collapsed, housing starts halved and land prices fell sharply — does that remind you of more recent periods, especially 2008?

Well, 35 years ago that led to many building firms falling by the wayside although it also was the time when Barratt rose to a size equalling hitherto-unrivalled Wimpey.

There then followed another growth spurt: by 1980 more than 14 firms were building 2,000 or more homes a year, while new national housing companies included Beazer, Ideal Homes, McLean, and the then-new retirement specialist McCarthy and Stone. In the early 1990s the next market crash led to more headlines that sound eerily familiar today — £2.5bn written off land holdings, many building firms run down or sold off, and the start of a decade of consolidation.

Mergers and acquisitions meant that by the Millennium there were three builders constructing over 10,000 homes each per year and the emergence of what Calcutt calls "the focussed house builder" — the firm that does little, if anything, apart from building and in many cases builds only homes rather than any wider portfolio of property types in the commercial, retail or industrial sectors.

That "focussed house builder" business model, when we look back on the recession of 2008 and 2009 and beyond, may have been the undoing of some builders.

The business model consists of various elements:

- an in-built underlying assumption, fuelled in recent years by government pre-occupation with targets, that there is an overall shortage of properties in the UK. This makes it possible (and in many areas, probable) that a high-density scheme will eventually get planning consent from a compliant planning authority. Although the consultation and permission processes are slow, the underlying assumption is that they are led by under-resourced and sometimes ill-trained planning officers at planning authority level

- that the builder's estimates for the density and pricing of units on a site are determined approximately at the time that land is purchased

- the builder then seeks to validate that density and pricing plan, and seeks to create immediate cash-flow, by selling off-plan units — sometimes years before construction work is scheduled to complete
- the builder seeks to directly or indirectly manage the project, and its costs, at all levels; that it manages land acquisition, surveying, design, sales, marketing and public relations.

When elements of this model collapsed in 2008 the residential establishment panicked. It sought government help to prop up the business model — perhaps an understandable short-term reaction — but what is a shame is the slowness with which it showed any sign of analysing whether the model itself required refinement or replacement.

For example, Stephen Conway of Galliard Homes — which like many developers had not in the past spoken out in favour of government intervention in the marketplace — advocated banks being obliged (by law if necessary) to redraft their loans to builders.

"At present the UK's major house builders are being forced to realise cash in order to comply with their loan covenants. The only way they can do this is to dispose of their existing stock of finished houses which they are doing at almost any price in order to avoid possible debt-for-equity swaps or expensive covenant renegotiations. Another increasingly significant component of the current housing market is the resale of mortgage repossessions at prices that, by their very nature, do not reflect their true value", he said in late 2008.

He added:

> In addition to this and in response to the new Council of Mortgage Lenders regulations, valuers have no choice but to base their valuations on these depressed prices with no allowance made for factors such as new build premiums or the discounts inherent in a forced sale either by a builder desperate to sell or by a mortgagee in possession.

His point was that pressure was being put on vendors to sell at reduced prices that in turn depress values that increase the pressure on the sellers — and so on. This downward spiral in values not only crippled the housing industry, he said, but also forced banks to make provisions that they normally would not consider.

Conway added:

> What is needed is some intervention to break this spiral. I believe that if the major clearers [banks] were to take a longer term view on housing

> stocks by easing the covenants on the quoted house building sector and allowing them a reasonable time to dispose of surplus stock this would go a long way to easing the situation. It must be in everyone's interests for the Government to support the clearing banks who, in turn, must support the housebuilding industry by allowing an ordered sale of surplus stocks over a reasonable period at reasonable prices. One way of achieving this may be for the Government to provide a vehicle to acquire the current surplus for rental to the private sector in the interim. This stock could then be sold as and when the market allows.

The Home Builders' Federation (HBF) was similarly short-term, asking for three actions. First, it wanted the Government to order banks to return to "sensible levels of mortgage lending"; second, to reallocate the Housing Corporation's budget to buy empty homes and unlock sites "on which development could start immediately with an injection of up-front public money"; and third, to encourage new housing demand such as allowing self-invested personal pensions to invest in residential property.

"We've reached the stage where radical and decisive action is needed to assist the housing market", said Stewart Baseley, Executive Chairman of the HBF at the end of 2008.

Two points emerge here.

The most obvious is that developers did a remarkable about face: they had been at best sceptical and at worst openly opposed to what they called unreasonable government intervention in the market at all levels, from Home Information Packs to increased regulation of building methods, health and safety legislation, and greater environmental protection. Yet here they were, demanding government action, basically to save their necks.

The less obvious point, but tied in with the first, is that there appears to have been no period of retrospection about whether their business model was wrong. This is unsurprising in an industry fighting for survival but no developer even since has made a substantial comment doubting the wisdom, for example, of concentrating high levels of building in the same city centres — indeed, that policy led to the glut of flats that by 2009 the same developers wanted the Government to buy from them.

So is this process an appropriate business model for the future?

Problems with the existing model 1 — too many apartments

Many industry insiders think not, starting with the types of properties that developers intend to build.

"The development-led model will shift from flats to houses. There's a massive problem in the interim before that happens — what to do with land allocated to flats in city centres and for more general residential development on browfield sites? Some of that will be worth almost zero now and you have to ask, who'll buy that?" questions Liam Bailey, Head of Residential Research at Knight Frank.

The glut was not down to government targets, he says.

> For sure, local and national government policies allowed it and possibly even encouraged it, but there's no doubt that the driving force was that developers saw flats and more flats as the quickest way to make the maximum amount of money.

The concept of city centre living started off as a great idea but over time it became bastardised. It became a speculative market and there began to be some very poor design — far worse than you'd get in a social housing development for example. Some of the cladding on these blocks already looks pretty ropey. Some of the homes are crap, not to put too fine a point on it. Now the idea is seen ultimately as a failure.

By the time the building hiatus is over, the price of land landscape may easily be a third down on what it was even in late 2008 and future development appraisals will have to be based on that, says Bailey.

"One of the biggest problems is that land owners won't sell at the current much lower prices. They just won't accept that a site which a while ago may have been worth £10m for residential development is now worth £3m. So in terms of targets and potential new schemes, one obstacle will be the refusal of land owners to face up to reality" he warns.

Problems with the existing model 2 — the homes are too small

"With very few exceptions [once existing projects have been completed] I think there will be no two bedroom or one bedroom apartments built in Britain in the next 10 years. But the problem that throws up is that there's almost no 'easy' developable brownfield land left — what remains is expensive and difficult and contaminated so there will be immense pressure for Greenfield development", says Jon Neale of Knight Frank's residential development research team.

"In terms of types of homes, developers are pretty conservative. If they're now wanting to build family homes, expect more of the executive home types, perhaps of a higher density. They're not bad in themselves but they're not the finest design ever — however, they're tried and tested, were relatively popular, and can be resuscitated, so they aren't without merit" he says.

"In terms of regeneration, obviously many schemes will have to be reconfigured because they were based on flats. Some schemes might fall by the wayside but good, experienced developers will be able to see a way through" suggests Neale.

Sizes will have to change, too.

It has been well-documented that British new-build homes are the smallest in Europe. The average floor space of a new home here (so including houses as well as flats) is just 818 square feet; Denmark is highest with 1,475 square feet and Greece comes second with 1,361 square feet. In France it is well over 1,200 square feet.

The HBF and other property industry data show that a typical new home built in Britain this year will be some 55% smaller than one built 80 years ago. And there are now 20% more rooms than a quarter of a century ago, meaning we have moved from a country of few large rooms to one of numerous small ones.

Four reasons lie behind this trend.

First, unlike most nations, Britain has no legal minimum space standard for homes.

Second, many schemes of flats built since 1999 have included smaller-than-usual studio, one- and two-bedroom units — these have been aimed at landlord buyers who have assumed tenants would make do with small rooms.

Third, high land prices have obliged developers to pack in vast numbers of tiny homes to try to make a scheme make sense, financially.

Finally, the Government's ever-denser build targets have effectively forced local council planners to permit ever-smaller homes.

This trend has not been without some public support. "In recent years people have asked for home offices which tend to be small. And, until now anyway, they've eaten out more, so some kitchens are smaller. There may now be more small rooms where there were fewer large ones", says Chris Fayers of Westcountry developer Eagle One.

"Remember design is as vital as space. I've known well designed smaller apartments that appear and operate in a bigger and better way than a poorly-designed larger one", claims Julian d'Arcy of the Leeds office of estate agency Knight Frank.

In addition, small rooms are cheaper to heat. "They may give tenants and owners more flexibility, to heat some areas and not others, optimising their fuel bill", says a spokeswoman for the Citizens' Advice Bureau.

But in the long term, rooms in new houses will almost certainly get larger.

Some politicians — notably London Mayor Boris Johnson, who says the Capital's new small homes are "shameful" — want a return to the so-called Parker Morris standards, introduced for council housing in 1967. These say there should be 355 square feet of internal space for the first occupant with each additional resident getting another 140 square feet. So a one-bed flat for two people should be at least 495 square feet.

Contrast that minimum with a converted studio flat on sale near Gloucester Road in central London on sale in early 2009. It has just 170 square feet and costs a cool £225,000 even in today's declining market. Even if you accept a studio flat has only one resident — and many actually have two — this example is under 50% of the standard promoted for social housing 41 years ago by Parker Morris.

The smallest homes are most routinely found in new-build schemes. Barratt, the troubled developer, has built what some observers believe may be the smallest new homes in the lead up to the recession — "Manhattan Pods" in Essex, which are each just 366 square feet.

Developers, unsurprisingly, are not happy at the prospect. The Chairman of Countryside Properties, Alan Cherry, has already gone on record claiming that any insistence on larger rooms "inhibits designers to some extent."

But anyone who has lived in a tiny new home or a converted flat may well look kindly on the credit crunch inadvertently producing a more expansive lifestyle. And if they have a cat, they may at last have room to swing it.

Problems with the existing model 3 — build quality is not consistently high enough

A report by the Office of Fair Trading (OFT), the Government watchdog supervising business practices, reveals the widespread number and range of problems suffered by people who purchase new homes.

A third of buyers suffer completion delays and many have to leave their old home, rent temporarily and wait for builders to finish their homes later than promised. Then some 70% of buyers find faults — sometimes more than 50 faults in one unit. Some 10% of buyers say their new homes are "poor" or "very poor".

It doesn't end there. Other statistics are equally dispiriting for the residential construction industry.

1. Numbers of buyers finding faults or problems with their home rises to 76% in the case of those purchasing apartments rather than houses.

2. Some 63% of those finding faults reported between one and 10 faults, but 5%of them claimed there were 50 or more faults.

3. Most problems concerned allegedly poor decoration (29%) or plasterwork (25%), issues with glazing and windows (27%), faulty central heating, hot water, or kitchen appliances and units (25%), problems with internal doors (24%) and faulty or missing electrical sockets (20%).

4. Half of those experiencing problems said they had been resolved in two weeks; another 22% said they were rectified between two and four weeks later; but 5% had to wait six months and 2% claimed they had to wait a year for a satisfactory repair;

5. Over a quarter of buyers who found faults or problems felt that homebuyers had "little or no protection" when buying new homes, despite warranties that allegedly promise swift repairs soon after a purchaser moves in.

This isn't the first boot up the backside for builders. Back in 2004 the Barker Report — at that time, the most comprehensive survey of the housing industry ever undertaken in the UK — criticised developers' build-quality and after-sales service.

Their response — surprisingly similar in almost every case — was to introduce their own in-house survey which with Soviet-style predictability would produce 90%-plus approval ratings from purchasers. The 2008 OFT report shows what many people might believe to be a more realistic appraisal of the industry's performance.

If the latest figures were from car buyers or restaurant visitors there would be uproar. Luckily for the residential industry, the OFT statistics were released over the summer of 2008, when the media were pre-occupied with market gloom.

Problems with the existing model 4 — they are expensive compared to other alternative accommodation types

Richard Donnell of property consultancy Hometrack has analysed the comparative average costs of new-build two-bedroom apartments in 10 key areas around the UK, contrasted with the average costs of a second-hand two-bedroom flat. The results show what Donnell calls "a massive disconnect in local markets", whereby new apartments are considered more expensive than comparable second-hand properties and of course vastly dearer than larger older homes:

Location	New-build two-bed flat	Second-hand two-bed flat
Croydon	£252,439	£170,000
Hounslow	£239,395	£182,000
Islington	£536,092	£300,000
Westminster	£539,500	£418,000
Manchester (city)	£174,823	£120,000
Liverpool	£151,934	£105,000
Sheffield	£144,943	£95,000
Newcastle-u-Tyne	£120,360	£94,000
Birmingham	£172,411	£95,000
Leeds	£152,491	£107,500
Nottingham	£101,500	£77,500
Bristol	£157,862	£145,000
Reading	£210,431	£155,000
Southampton	£161,098	£122,000
Cardiff	£152,768	£115,000

Source: Hometrack

Of course, developers are not pricing their properties to be uncompetitive but because of high land prices and (as we shall see below) an expectation of capital appreciation, so these builders have pushed the envelope on prices, time and time again.

Towards a new business model

Now of course developers are not to blame for the high land values of the 1990s and 2000s in particular, which created the need for more dense developments of small properties at premium prices to make the figures add up. But now that land prices are plummeting, and demand for apartments so clearly on the wane, and buy-to-let on the back foot for some years to come, this is surely the time to rethink the business model.

Some developers believe that, for example — just as with estate agents who relied entirely on sales and were thus overwhelmingly exposed when transaction numbers collapsed — so residential developers who did not diversify in the good years were far weaker to the market turndown when it came. A few of them were already modifying their business models and this may be a pointer to how others should act now.

"We had suffered the pain before and were determined not to be caught out again. We put in measures after the previous recession in the early 1990s" explains John Hunter, founding partner and Chief Executive of Northacre plc, the upmarket London-based developer.

"Firstly we chose not to concentrate solely on development, such that the proportion of our income from construction development now is only 30% whereas 50% of our income comes from interior design, and the remaining 20% from management fees, consultancy and architecture" says Hunter. To demonstrate the size of the shift — "an entirely deliberate one" says Hunter — in 2004, interiors accounted for just 20% of his firm's income.

"Secondly we sourced business and income from all over the world. We opened an office in the Gulf, which is an area of the world where you need to be and which will be a bigger part of our work" explains Hunter.

He says diversification can be built up through outsourcing or, as he himself prefers, by keeping it in-house. "They're both perfectly good models so long as you diversify, which is the important thing."

He also calls, unashamedly, for developers to return to what he says are "simple basics".

"There was a time when you bought a site, built homes and sold them. That needed high margins to accommodate any problems and good planning to try to avoid those problems happening. But as a developer you couldn't rely on capital appreciation to generate profits and you couldn't rely on off-plan sales to generate cashflow", he says.

> We've moved to a model reliant on vast turnover and tiny margins, and that's what went wrong in 2007 and 2008. Our whole industry now needs a cap on the amount of external financing we get. There should be a maximum of 75% of a property's price or a scheme's development costs that can be borrowed. That's where it ends if we're serious about who builds and how they build.

Instead of that, the corrupted existing model encouraged too many developers to pay too much for land, assuming capital appreciation of the units — fuelled by off-plan sales — would cover the price.

> With residential development you can in theory reduce your unit pricing so those units can sell, but you can't do that if you have small margins and if you've paid too much for the land to begin with because you've banked on appreciation to see the project into the black. That's what's wrong with the old model.

Rival developer Andrew Murray, who runs niche sustainable building firm Morpheus, agrees.

> There's got to be more diversity in the residential sector. We may buy and sell like we do now but there's got to be much stricter risk assessments by those who fund the developers. In recent years it's been amateurish and a bit of a con, like self-certification. A bank has asked a developer how much a scheme is worth, the developer says 'x' and the bank then says 'OK'.

Simon Wright of the eponymous Simon Wright Homes was another developer to see the writing on the wall.

In late 2008 he said:

> After a meeting I attended with the Bank of England back in August 2007, I saw signs that the market was on its way down, at which point I sold off a lot of land, mostly to housing associations, thus minimising the company's exposure to risk. Simon Wright Homes is not just a standard 'house builder' — we have several strings to our bow, including a dry lining business which is the biggest in the south east. In the last year we have diversified further, setting up several businesses under the Wright

> Group umbrella including SJW Civil Engineering, Wright Construction UAE — which potentially has several major projects in the Middle East — and Wright Plant Hire. Because of our in-house construction skills, we can over a turn-key service, and as a result, have been awarded a number of construction projects, including schemes for Housing Associations who bought land from us. Diversification has been key to the success of our business therefore I believe it is vital to the survival of the industry.

The same can apply to bigger companies. Back in 2004, Berkeley Homes showed what could be done.

The astute, if not exactly self-effacing, managing director Tony Pidgeley quit the mainstream top-end residential marketplace of central London and moved instead to mixed-use schemes in city centre regeneration areas, saying:

> Berkeley is placed in a uniquely strong position to lead the continuing revival of Britain's cities and the retrieval for housing of derelict land in urban areas.

He has had his share of hard-to-sell apartments during the downturn but nonetheless his company failed to attract the depressing headlines that so many residential-only volume builders received.

Indeed his firm's latest figures — up to November 2008 — add up to what Pidgley describes as a "robust set of results" despite a 60% slump in sales, and show the wisdom of planning for a highly cyclical market — and, by the way, not borrowing much.

"We're concentrating on generating free cash flow and protecting the balance sheet. This builds the strength to withstand the more challenging times and to then take advantage of the opportunities that will follow. While I'm pleased by these results, it's the cash generation of £140m that is the outstanding feature of the performance" he says.

Remarkably, and in almost complete contrast to most of those who relied entirely on off-plan sales and assumed a rising market, Berkeley's balance sheet shows no write-downs, no debt whatsoever, that net cash generation of £140m with £810m of cash expected from forward sales — at least half expected from investors, even in the new volatile market.

"This is in our normal range for investors who continue to be influenced by the lack of alternative investments, with doubts over pensions and the stock market and with low interest rates producing unattractive returns on interest-linked investments" says Pidgley.

This strategy has not only seen Berkeley weather the storm of the downturn, but be in a good position to take advantage of the coming upturn.

Other alternatives to the existing way of doing things

We can see that alternative business models are emerging.

One is the traditional-style development company with a more conservative approach than in recent years to land acquisition, less reliance on off-plan sales and ever-rising sales prices, and aiming more at family and owner-occupier purchasers rather than investment buyers. The developer may become more diverse, operating in markets other than purely construction, other than just in the UK, and possibly other than just residential.

But there are other models too, as the Calcutt Review suggested.

One alternative that is already gathering pace is the so-called Investor model. This is when a builder retains directly, or through a third party, a long-term interest in a developed site by delivering a range of management and other services to the site (renting properties on its own behalf or for private buyers, or looking after communal facilities are two examples) and in return get a continuing income stream not reliant on new sales.

Calcutt says under this model, yields may be smaller because "the developer trades a proportion of the up-front development profit for the opportunity of long-term revenues plus future capital growth." A few months after the Calcutt report was published, many developers — unable to sell their homes in recessionary Britain — instead rented them out; some considered adapting their schemes to allow this to happen in the long-term, putting into practice what this "investor model" preaches.

One further alternative is the Registered Social Landlord (RSL) model, whereby RSLs develop homes for sale (some, but not all, through one of the many shared ownership schemes that exist). Calcutt obviously believed this model could lead to an increased share of the new homes total being taken by housing associations, once the slowdown ends.

It is something of an indictment of private developers that Calcutt should say that RSLs are "much more likely to focus on quality and sustainability in design and construction, and to welcome innovation" than many existing private developers, although the review noted that housing associations would not be able to achieve the completion numbers of their private rivals. This RSL alternative will only work if a wider range of property types and a generally higher build-quality characterise the residential development industry in the 2010s.

Another alternative business model — for the industry as a whole, perhaps, rather than any single developer — is to encourage more self-build.

Far from giving in to the sneers of the property establishment, which has long mocked the 15,000 homes built this way in the UK each year (either directly by individuals, or by those who contract architects and builder), Calcutt believes the self-build sector has more to offer:

> Advocates of self-build argue that, with more support from Government and local planning authorities, the volume of self-build could grow significantly. It certainly has too large a share of the market to ignore. For obvious reasons, self-build developments characteristically achieve higher quality of specification and better cost-in-use. They are more likely to be innovative, not least in sustainability (although self-builders tend to prefer traditional building forms). Self-build exploits small sites particularly well, and is more likely to provide opportunities for smaller local builders — the larger house builders tend not to be interested in self-build, which delivers lower though less risky profits than their investors expect.

The report concludes:

> Self-build on its own cannot deliver the increased volumes of house building that are required but it has a contribution to make. Central and local government housing policies tend to neglect self-build, which we think is a wasted opportunity. We recommend that Government and its agencies disposing of land should consider the opportunity for self-build and should aim to offer a proportion of the land in the form of small plots, where possible with ready access to services and other infrastructure for sale to self-builders. Local planning authorities drawing up their strategic housing land assessments ... should similarly aim to identify a supply of small plots suitable for self-build and other smaller house builders.Now no one in their right mind says the collective might of the self-builder or the ambitious RSL will rival the volume house builders, but one thing is worth thinking about — whether we like it or not, both housing associations and do-it-yourselfers are expected, according to Calcutt, to take a bigger share of the new-build cake in the future. That means private developers will have to take a smaller share.

If councils are given increased funds and powers to resume having their own direct labour organisations and new-build programmes — notwithstanding the stigma that some may associate will local authority housing — then again a slice of the cake will be taken away from private developers.

Failure to act now, refusal to adopt a new business model and attune to the demographic needs rather than maximum short-term profits, may see the volume builders lose much more than a year or two of sales in a slowdown.

It could mean a future where they are permanently smaller players on the British housing landscape.

6 The Future — Everyone Raising Their Game

No one should underestimate the difficulty for estate agents and developers in reviewing their current business models and revising them, or abandoning them completely in favour of more sustainable, modern alternatives.

For the publicly-quoted companies in particular, with shareholders to satisfy at short-term intervals, the task will be complex and sensitive. But shareholders, business partners and all others involved in the property industry must surely see that after the "cull" of participants in 2008 and 2009 there is a need for change.

How long have players got to undertake this transformation before we all return to the sort of transaction levels, less restrictive mortgage market, and buoyant prices we all enjoyed for more than a decade before the slowdown?

Too many commentators have had their fingers burned over the past two years to expect precise predictions for the future. Besides, history suggests Britain's latest recession will run for some time. The recession that began in 1980 lasted five quarters and saw output drop by 4.6%; a decade later there was a recession of similar duration but with a 2.5% fall in gross domestic product. In 1991 growth resumed for a short period but there was a further negative quarter in 1992. These things do not end quickly or neatly.

"Forecasting models are notoriously wrong but when factors in the model include likely future Libor spread and the creditworthiness of mortgage providers and the country providing the financial guarantees ... the process becomes an almost hopeless task" says Knight Frank's research head, Liam Bailey.

Hopeless, perhaps, but there is at least some consensus that the recovery from the housing downturn will be slow and patchy. Most commentators who have dared to make predictions say that any return to normal may involve a definition of "normal" which will be quite different to that which existed in, say, 2006 or 2007.

How long?

Savills was the first major estate agency to predict the downturn — "and we took some shit for doing so in the first couple of months of 2008", admits one of its research directors, Lucien Cook — and now it is one of the few to try to forecast in detail what may happen in the years to come.

It says house prices in London and the south east of England will improve in 2011 although they will not return to the "pre-slump" levels until around 2013. The south west, the east and west Midlands and the East of England will follow by about 2014. Then a year later there will be recovery for Wales, the north west of England and Yorkshire. Northern Ireland and north east England will be the last to recover around 2016 or even beyond.

"In London and the south east in particular, the underlying drivers will kick in again when mortgage lenders are once again able to operate in more typical conditions. The end of the credit crunch is the key factor here. The financial markets appear to be pencilling it in for 2010, at a time when economic growth is forecast to return to more than 2% per annum", says Yolande Barnes, another Savills research director.

"Lack of affordability will not be a constraint when the credit crunch ends, but the lack of purchasing power amongst a growing proportion of the population will be. Many would-be home owners are currently being 'priced in' to the market but many existing owners are experiencing negative equity for the same reasons. Irrespective of the current market malaise, many more will still find themselves priced out in the years ahead, especially those non-home owners unable to afford deposits", she says.

The simple shortage of property — the mismatch between long-term supply and demand — will keep prices rising over the long term, she says.

The recovery will be less long-winded according to the Centre for Economic and Business Research (CEBR), a consultancy that has been amongst the more level-headed of the non-housing commentators over recent years.

It says with completions of new homes likely to remain massively below government targets until 2013 at least, prices will ultimately rise again before that time.

"The credit crunch has caused a shock in the housing market much bigger than most people expected and we are now seeing the second round effects of falling confidence and a slowing economy. When prices have fallen in the past we have seen house building slow quite rapidly but take a lot longer to come back, which leads to demand outstripping supply. With the fundamentals of the housing market still relatively tight, the credit crunch might already have sown the seeds of the next house price boom", according to CEBR spokesman Richard Snook.

Ben Read, Senior Economist at CEBR claims the secret may lie with the Bank of England. He says:

> The prospect of several interest rate cuts ... combined with some easing of mortgage rationing, is likely to lead to demand starting to pick up. The sharp drop in building completions that we are currently seeing is likely to continue, and this may mean that prices recover more quickly than people imagine.

Over the longer term, the Government will be concerned that with every year that passes we get further away from house building targets. This inability of the supply side to respond quickly to increasing prices is an ongoing problem which means that prices are able to rise quickly in the short term, and that over the long term prices will generally outstrip growth in incomes.

Whither targets — and bureaucracy?

Back in 2007 Gordon Brown announced that in England alone there should be between 240,000 to 297,700 homes built annually until 2016. In total between 2.9 million and 3.5 million new homes should be built by 2020. In 2007 — just before the downturn hit the new-build market — some 174,900 homes were completed; this was still below the target but on an upward path from previous years.

Yet in 2008 it appears just 110,000 were built with many forecasters suggesting that the following two years would be down to a miserly 60,000 or so per annum.

Jim Ward, another Savills research team member, predicts new-build levels in England will not return even to 150,000 per year — scarcely half the Government's target — by 2013. He says:

> Once sites are moth-balled there's inertia in the system as it takes time to rebuild teams and return to the same master-planning position on more complex sites.

To exacerbate the shortfall, some advisors to Gordon Brown — chiefly the National Housing and Planning Advice Unit — says the original targets set by the UK Government were inadequate anyway and should be up to 10% higher because they systematically underestimate the shortfall in future years.

The slump in building has not only affected the private sector, but the public one too. Many affordable homes for sale, shared-ownership or rent to key workers and the low paid are built as a result of so-called section 106 deals — the planning jargon that means developers have to build a number of affordable homes in return for agreement to build private sector homes.

The Thames Gateway is a good example of just how the slowdown in 2008/09 can have long-term consequences which widen the mismatch between supply and demand for homes, especially in the social housing sector.

With little private sector building going on in the Gateway until the slowdown ends, the section 106s are not happening and this is likely to remain the case for some time. "Intelligence suggests some developers are stepping out of the market [for] anything up to two years", says a spokesman for the London Thames Gateway Development Corporation, an Urban Development Corporation in the area.

Meanwhile, those relatively few developers still willing to build will insist on fewer section 106 obligations.

"Developments that received permission recently were based on significantly higher market values. S106s will need to be renegotiated" warns Jim Briscoe, Affordable Housing Chief at CB Richard Ellis. "The old deals were predicated on private houses and land values rising. They took years to achieve and now they're almost worthless" says a blunter Jon Neale, Head of Development Research at Knight Frank.

Renegotiation is not a simple task given the vast number of bodies — "partners" as the Government calls them — that oversee the Thames Gateway, arguably the flagship location where British new home building is showcased.

There is the Department of Communities and Local Government lead department, plus seven Local Regeneration Partnerships and two Urban Development Corporations. Add to the mix three Sub-Regional Partnerships plus the Thames Gateway Strategic Partnership which links all of the groups above. And, of course, there are the formal consultation processes which exist with commercial and residential developers, Registered Social Landlords, private and public sector bodies, utility companies and local communities.

The Thames Gateway is not typical of the rest of the UK — for which we should be thankful — but it demonstrates how the plethora of planning bodies which exist in almost every part of the country will make any progress on meeting new home targets very difficult even when financial conditions improve.

"It's a mess" says Neale of the Gateway situation. He contrasts the slow progress in that area today with the decisive role of the London Docklands Development Corporation between 1981 and 1998:

> There have been successes in the Gateway, notably the [Land Securities-masterplanned] Ebbsfleet area, but there's not a lot else to boast about.

Individual bodies say the Department of Communities and Local Government (DCLG) is the only forum that can monitor Gateway-wide progress. But a DCLG spokesman admits "there are loads of numbers flying around" and he cannot give a reliable figure for recent house completions.

The DCLG also says neither it nor any other Gateway body has a system in place for assisting hard-pressed local authorities with renegotiating section 106 deals. "It's about commercial confidentiality. We don't interfere, it's down to them" says a spokesman.

That fundamental mismatch of supply and demand worries not only estate agents and developers but bodies like the National Housing Federation (NHF).

A report published in 2008 by the NHF and researched by independent think tank Oxford Economics, suggests the average house price in England will rise 25% by 2013. NHF Chief Executive David Orr says:

> House prices will increase substantially over the mid to long term. Demand is going up, while the supply of new homes is going down. This means that as soon as the economic outlook improves, house prices will resume their previous upward trend.

So how will developers respond to all this?

Liam Bailey of Knight Frank says low land values and falling build costs will set the scene for a new phase of development activity, possibly from 2012 onwards.

"However by the time the recovery comes, traditional house builders may not necessarily own the land with planning permission — we believe a large portion will be in the hands of sovereign wealth funds, other institutions, local councils or wealthy individuals by that time. New approaches may emerge involving joint ventures or consortia developing new communities over the longer term. We may even see 'build-to-let' become a reality", he says.

But he warns that the picture is not one of inevitably greater levels of building as the construction industry regroups over the next five years. There are, he says, simply too many expectations on developers.

"The boom in land values and house prices led to a huge increase in planning obligations on developers. In many consents that are in place, call for new transport infrastructure, flood defences and, particularly, large amounts of affordable housing, much of which is no longer viable The government and local councils will have to adopt a more flexible approach if they want to see activity return to historic levels, let alone meeting their targets", he cautions.

Like his rival analysts at Savills, Bailey and Knight Frank see mainstream house price recovery starting in the capital and then trickling outwards.

"London will start its recovery first in 2011 but won't be back to its late-2007 pre-slump levels until 2015. Most of the rest of the UK will take much longer. In the case of Northern Ireland, full price recovery won't happen until 2019. Then we should be back to normal" says Bailey.

If you had a pound for every time property professionals talked of what may occur when markets/buyers/sellers/developers/investors "return to normal" you would have enough money to buy a bank (which of course these days, may not be that much).

The new normal

So what is this longer term mood music going to be like? What is the "normal" we want to return to?

As recently as 2007 we had some basic assumptions about "normal" life. They included a belief that people "naturally" wanted to own a home, and possibly more than one. There was a similar assumption that people liked house price rises because it made them feel more comfortable. Once people reached a modest level of wealth, surely the natural thing to do was buy a little villa in Spain? The middle classes may choose France instead, while the poshest looked to Chianti or the Caribbean. That seemed "normal" — at least if you had the money to do it.

If you planned for the future just a few years ago, wasn't it inevitable that you bought a flat or two to rent out, running them from the rental income but gleefully anticipating the day when you sold and pocketed all that lovely capital appreciation? That was not necessarily to fund a luxurious lifestyle, but to provide a decent pension in old age.

Well, maybe the next "normal" isn't going to be like that at all.

What we became used to in the period after the early 1990s recession and until the current downturn was, in fact, an atypical set of circumstances, and historians of the real estate industry in the UK insist that there have been "episodes" of 10 to 20 years at a time. Each episode has been distinctly different from the past and future.

Until the outbreak of the first World War, just under 9% of homes in the UK were owner-occupied — slower than some other parts of western Europe and far behind the 46% figure in the United States.

As the *Sunday Telegraph*'s Economics Editor Edmund Conway has written in the *New Statesman*:

> Even the wealthiest young men would prefer to take rooms —lodgings — rather than buying or renting their own properties when coming to London. In most circles it was perfectly normal never to own your own house. Things changed after the world wars, as successive governments embarked on policies to find 'homes fit for heroes'. Controls were imposed on landlords and millions of pounds were poured into homebuilding projects. Meanwhile, inequality was falling, meaning many more middle-class families were suddenly able to afford to buy a home.

New-found optimism in the 1950s was augmented by genuine wealth beginning to spread throughout society in the 1960s. Governments of all colours began to introduce tax breaks on mortgages and the

Thatcher revolution created the right for tenants to buy their council houses — a policy which, ironically, spread home ownership throughout the "sub-prime classes" and simultaneously saw the end of meaningful council house building and the demise of apprenticeship schemes which used local authority stock as top-class training grounds.

By the time Tony Blair entered Downing Street, Britain appeared a very different country to the rest of Europe. By then it had some 70% of households that were owner-occupied — in France and Germany first-time buyers were much older because, as in Britain in the 19th century, it remained perfectly acceptable to rent and many people spurned the shackles of mortgaged property all of their lives.

Is that mainland European model a sign of what might happen in Britain, post-recession?

"We believe that a wholesale shift towards a French or German model, where long-term or even lifetime renting is considered normal, is a step too far. But we do feel that occupation options will become more flexible in the future with people able to choose from a mix of shared ownership and rental options" says Liam Bailey.

The rental sector will increase, he says.

> A more cautious approach among mortgage providers could mean this is the only option for those who cannot access social housing. A larger rental sector will emerge from this credit crunch by default as lenders opt to rent out repossessed houses instead of selling them on, for a potential loss. The yields and cash flows from such portfolios will prove attractive.

And those new homes that are built will not necessarily be still more apartments — thank god — and will certainly not be primarily aimed at investment buyers.

"Those developers left in business will no longer focus on high-volume inner-city flat developments. They will try where possible to target family houses in the outer suburbs where there is more tangible demand. Urban development will concentrate on 'edge of centre' brownfield sites where there is a possibility of developing high-value mixed accommodation as part of a wider gentrification", predicts Bailey.

So what will the future look like? Experts vary. Some think it will be new, exciting, and every bit as buzzy as the 1980s promised; most, however, think it will be like the 1970s albeit with the increased choice that we consumers have enjoyed in more recent decades.

1. **Peter Rollings of Marsh & Parsons estate agency**:

The new normal will initially be a much quieter market with nothing like so many speculators. But it's my view that as in all cycles this will gradually increase within the next five to 10 years and we'll almost certainly see a boom of sorts within the next decade. The UK and London especially are 'small' and with population set to increase to 70 million in the coming 20 years, there will quite simply not be enough houses. This may mean that the private renting sector will continue to increase and my view is buy to let will still remain extremely strong. It will become increasingly common to have a second property for vast numbers who will have lost faith in faceless fund managers investing their hard earned pension contributions in an 'emerging market' somewhere in the world and will instead invest in something real and tangible. Prices will start to recover slowly; however with a total drop of 20% to 30% ownership will have been opened up to a lot more people — every cloud has a silver lining. Aspirational people will of course still wish to buy second homes, while smaller flats within London will probably increase in value as commuting becomes more expensive and more crowded and this will continue to grow.

2. **Ed Mead, Douglas & Gordon estate agency**:

The days of high owner occupation will dwindle and the stigma attached to renting will disappear. Buy to let will flourish but there are many very wealthy organizations and individuals who will replace the long term 'sitting tenant'. At last you'll see the demise of crappy developers who literally got lucky off the back of a rising market. Any idiot can make money then but the trick is to be clever and design led, and it's those clever developers who'll remain. Second homes will continue but destruction of communities will most likely lead to the government regulating to curb 'out of towners'. So 'normal' will be a return to pre-1978/1979 purchasing power and a return to KYC — that's old fashioned City speak for Know Your Customer ... and about bloody time too.

3. **Barry Manners, Director of Chard estate agency**:

People will save money and spend what they have earned and saved. The property casino where everyone was a developer — whether they worked in the media, arts or as a refuse collector — is probably over. Buy to let will continue because people working in London will no longer see the imperative of buying. I suspect yields on smaller flats will actually improve. Larger family houses that cannot be let to sharers will see yields fall — massively. People will want second homes, hopefully because they want to spend some times with their families in them rather than treat them as a commodity or pension. This will mean they may once again become the preserve of the established wealthy.

4. **Martin Lamb, Savills estate agency**:

We're going to be back to where we were in the 1970s. There will be first time buyers but they'll have to stump up 25% deposits. There will be investment buyers but they'll want good quality properties not desperate apartments, and they'll look to be in for the long haul not to turn a quick profit. But most people won't move house unless they really have to. This had to happen.

5. **Henry Pryor, housing market commentator**:

Normal might be 850,000 sales per year, down from 1.2m in 2006, 1.0m in 2007 and 650,000 in 2008. Values will be at 2002 levels which will mean a fall of 50% for many from their peaks. Buy-to-let will not return until 2011. Income in the form of rents will stabilise but capital appreciation will not, at least until 2012. The number of agent offices will have fallen from 14,000 to 7,000. Average commissions will have fallen from 1.4% to around 1.0%. Home Information Packs will have been scrapped although Energy Performance Certificates will continue for both sales and lettings. Typical mortgage Loan-to-Values will be around 90%. Development land will return to about 40% of its 2006 values. Commercial leases will change and upward-only rent reviews will go. Paying rent quarterly in advance will be replaced by monthly in advance. Roughly 40% of agents' high street offices will go as agents retreat to first floor offices. Local papers will have suffered as agents pull their print advertising. Pay-to-list portals like Rightmove will have been overtaken by free-to-list sites like Globrix. Estate agents will be licensed.

6. **Tim Wright, Residential Partner, King Sturge**:

The future residential market suggests a far higher level of rented property, especially if we move into an era of deflation. Cash savings rather than buying a house as an investment will become far more popular. Funding of developments and developers will change and the fall out will be felt more significantly in the mega or super sized regeneration project that are likely to be put on hold for some considerable time to come. The market will find a new level with house builders building smaller sites and mainly houses. The average age of first time purchases is likely to rise and some may never own their own house or flat. Social housing will remain a problem but it is likely central

Government will finance far more social housing, although not to the level of the 1960s when 100,000 units a year were built with public funds.

7. **Michael Cunnington, Director, MJC Associates (Mallorcan estate agency)**:

The major issue with overseas owners is that most of them are fairly well heeled and quite prepared to wait for the market to come back to 'normality'. It is well known that many overseas properties stay on the market for very much longer than the UK norm — maybe even two years or so — especially when overpriced in the first place. This time, however, it will take longer for the market to climb back to the original optimistic price level.

8. **Melanie Bien, Director, Savills Private Finance (mortgage broker)**:

Buy-to-let will not disappear but there will be fewer novice landlords buying a one-off property to rent out. Instead we'll see the savvy landlords return to the market, picking up bargains. Accidental landlords — those who have to rent out their property because they can't sell it — will decline in number as the market picks up again and they can sell once more. The case for home ownership has been severely dented and those who didn't experience the negative equity of the early 90s have now had their own lesson which will depress the market for a few years. There will be fewer mortgage brokers, lenders and estate agents, but advice will be more important than ever because there will be fewer mortgage deals available.

9. **David Pretty, Chairman, New Homes Marketing Board**:

While house builders would love to spread the risk and build more houses than flats, the reality is that demographic trends and planning guidelines have encouraged more flats. Planning guidelines need to be changed to tackle the current oversupply of flats, but I don't believe a dramatic change is needed to the overall mix and nor do I think it will happen because of those demographic and affordability factors. Off-plan buying won't be such a market feature in the short-term, as it is a symptom of demand exceeding supply. But that pent-up demand is increasing all the time, and when the recovery comes, so off-plan buying will return, particularly to the most popular developments.

10. **Andrew Murray, Director, Morpheus Developments**:

The mid- to lower-ends of the new build market will return to where they are now, eventually. Ironically labour is cheaper after a recession so those who are still building at the end of the slowdown will be returning in a much stronger position than before. Risk assessments by lenders and developers themselves will be much stricter of course. London's lucky because of its international interest, so the bounce back will happen there first.

11. **Tony Douse, Chairman, Environ Communities**:

'Normal' will be that the purchaser will not assume that house price inflation will provide their pension or equity for a house abroad. Maybe people will get back to thinking that their house is a home and not a money making machine. Perhaps people will think far more deeply in terms of the product as their stay in the house may be of much longer duration than previously.

12. **David Cowans, Chief Executive, Places For People (social housing developer)**:

Research suggests that households will have less time available for home related tasks. Many of these households would be in a position to buy ancillary services that fit around their lifestyles. Services such as landscaping, repairs and maintenance that are available through one port of call would appeal to a growing market. We are working on how we might use our existing facilities and external providers to offer a suite of services to residents.

13. **Alan Cherry, Chairman, Countryside Properties**:

I'm not so sure we will ever get back to what was until recently regarded as 'normal': I believe that housing markets will eventually become more active, but we will first need to see house prices stabilise. This will not happen until banks and building societies can increase the money they make available for home loans. Even when this happens I expect that lending criteria will be much stricter than in recent years. I hope all this will result in housing markets being more stable and house prices less volatile. Housebuilders will need to adjust to this changed situation and rely less on house prices inflation to make their profits. Housebuilding will be much more challenging. The need to adjust to the sustainability agenda will be important, as well as the need to produce better designed homes and better planned places will be vital if they want to attract buyers. The planning system will also be more demanding and I do not expect poorly planned and designed schemes to receive planning consent in the future. The prospects for the more responsible developers will be considerable. The need and demand for more new homes is substantial, so there is much to look forward to for those that produce new housing of the right quality that helps create more sustainable communities.

Final lessons

There is no point living a life of denial about the causes of the crisis. Most came from outside the property industry, of course, but some came from inside too, as we have examined.

In the future the entire industry — agents, developers, valuers, mortgage advisors and even property writers — must learn some lessons from what we have been through. We all have to raise our game, recognising that the world has changed and we have to work harder to achieve what we could so easily get when we lived in a booming market.

For example, property public relations officers and journalists have to recognise that we have shed the sloppy attitudes of the past. The public have fallen out of love with property, at least for a few years, so the mere hint of an investment opportunity or a successful scheme will not guarantee finding buyers or free publicity.

The portents are bad. In October 2008 Knight Frank's PR machine was ringing journalists to claim that new apartments at a West Country scheme were, in the agency's words, "flying off the shelves" thanks to owner-occupier and investor buyers. It sounded like old times, when off-plan sales were commonplace and people queued to buy new homes.

In reality, those calls came on the same day that the National House Building Council announced new home sales were 55% down on average and 80% down in the area where the development in question was being built. It was also the same day that the Royal Bank of Scotland began nationalisation talks with the Government. More ironic still, it was the same day that Knight Frank's own research department was preparing a document saying its own new home sales were down to 25% of their normal levels.

Clearly, not everyone at the company shared the same view about the market as the PR department. And of course a simple telephone call to the developer of that West Country scheme revealed that, in line with almost every other development in Britain, sales had in fact ground to a halt. What the PR came out with was, without putting too fine a point on it, complete nonsense.

But surely a sophisticated, successful firm like Knight Frank can recognise that the world has changed and simply trying to sell a property — whether as a story to a journalist or as a place to live to a buyer — needs something more subtle and more convincing than a wildly inaccurate claim that units are "flying off the shelves"?

This problem remains widespread throughout the property industry.

In December 2008 I received two e-mails carrying other press releases with the sort of get-rich-quick promise that was the meat-and-gravy of the residential property world until about a year earlier. It took very little research to find the promises were not quite as I had imagined them.

The first was from Cluttons Resorts, an estate agency selling resort villas and apartments at many of the world's top locations. The release carried the headline "Mauritius defies global property market trends" and spoke of properties for sale in a new phase of a development; some of those who had bought in an earlier phase were "already enjoying capital growth in excess of 30%", it claimed.

In the febrile global property market we live in, such claims sounded too good to be true.

On further investigation, I discovered from Cluttons that it had itself calculated that prices of homes bought early on had risen substantially (in some cases by well over 30% in fact). But as no home from that early phase had actually been sold the claimed appreciation was — as the company put it — "on paper".

Perhaps realising how this weakened the sales pitch, Cluttons went on to say:

> The fact no resales have taken place indicates the popularity of [the development] and that the majority of buyers are more interested in the lifestyle and quality of resort than making a quick return on their investment.

Quite so.

Of course there is almost certainly nothing directly misleading in Cluttons' claims, and anyway the absurdly-antiquated "buyer beware" rule means it is up to the purchaser, in the UK at least, to watch out for sharp sales techniques.

But I would be surprised if many members of the public would agree that one estate agent's untested claim about prices on one development actually justified the all-embracing claim that "Mauritius defies global property market trends". Something like "Estate agent hopes prices have risen despite global downturn" might have been more accurate, if less appealing to would-be buyers.

There is a similar approach taken in the second shrewdly-worded email from estate agency Chesterton about a Taylor Woodrow development on Spain's Costa del Sol that still has unsold units some years after they were first marketed.

A Chesterton spokesman is quoted in the press release saying that the Costa del Sol had been a popular holiday spot since the 19th century and received some eight million visitors a year "providing a great tourist rental market".

When I enquired, Chesterton declined to offer any rental market figures to back up its suggestion. Well, that's no surprise.

Anyone familiar with the Costa del Sol will know that the area is saturated with villas, apartments and houses to rent. Many are available to let during the "down time" when owners are not using the properties; others are permanently available to let while departing owners wait to find buyers — a wait which, for many vendors, takes years.

Therefore the rental market would in many people's eyes be described by words other than "great". But again the clever wording means the press release from Chesterton was not directly misleading — and it made no mention of how long the properties had remained unsold.

Naturally, estate agents and developers (and their public relations spinners) have a job to do, and that job is to shift units.

But the reputations of so many in the property industry have been damaged by shrewdly-worded suggestions of rapid gains in the past, that one would hope there would be more transparent and forthright descriptions in the future.

If there are, then that will raise the reputation of all of us in the property industry. If there are not, then with luck there will be journalists to make the facts a little clearer. But it would be much better if people just promised what could be achieved, and not made property into a 21st century snake oil.

But journalists can be as bad as PRs.

The magazine *Show House*, which promotes its parent company's house building awards and sales websites and sets itself up as a cheerleader for the new homes industry, carried a bizarrely irrelevant editorial in late 2008.

In the month that world stock markets were imploding and every UK house price index pointed downwards, and in the month that 30 niche builders went into administration, its editorial was about ... the alleged unfairness of government legislation regarding windows in new-build homes.

An important topic? For some, perhaps.

But it was hardly the most pressing problem for its industry readers facing threats to their business survival thanks to the market downturn and borrowing restrictions. *Show House* is a cheerleader for house builders but, again, shows that the industry needs a more sophisticated approach if it is to be taken seriously.

So everyone needs to get back to basics. That was the reason why a few senior property insiders played a straight bat in late 2007 and early 2008 when the market turned.

"Clients don't like to hear bad news and neither does a sales team but it's crucial to be honest and straightforward. Our research department ... was the first in 1992 to predict prices would rise by 20%. A lot of people thought at the time that was pretty punchy" says Rupert Sebag-Montefiore, Savills Chairman and Chief Executive.

He says the same philosophy was behind his firm's gloomier 2008 forecast, which was the first to tip-off the outside world that house prices were falling badly.

> When it became clear the markets were going to take a battering, we knew that the sooner our clients knew the severity of the situation the sooner they could take action, and the sooner the markets could recover. If we are going to be credible in the future we need to be trusted to tell it like it is.

Tell it like it is. Perhaps that may become the mantra of the estate agency and developer world that will come out of the recession. We can all at least hope so.

Index